D0214450

KANT'S THEORY OF SCIENCE

KANT'S THEORY OF SCIENCE

Gordon G. Brittan, Jr.

PRINCETON UNIVERSITY PRESS
PRINCETON, NEW JERSEY

Published by Princeton University Press, Princeton, New Jersey
In the United Kingdom: Princeton University Press, Guildford, Surrey

Library of Congress Cataloging in Publication Data will
be found on the last printed page of this book
Publication of this book has been aided by a grant from
The Andrew W. Mellon Foundation
This book has been composed in V.I.P. Bembo
Printed in the United States of America
by Princeton University Press, Princeton, New Jersey

for Vanessa

". . . to cognize anything a priori is to cognize it from its mere possibility. But the possibility of determinate natural things cannot be cognized from their mere concepts; from these concepts the possibility of the thought (that it does not contradict itself) can indeed be cognized, but not the possibility of the object as a natural thing, which can be given (as existing) outside of the thought. Therefore, in order to cognize the possibility of determinate natural things, and hence to cognize them a priori, there is further required that the intuition corresponding to the concept be given a priori, i.e., that the concept be constructed. Now, rational cognition through the construction of concepts is mathematical. A pure philosophy of nature in general, i.e., one that only investigates what constitutes the concept of a nature in general, may indeed be possible without mathematics; but a pure doctrine of nature concerning determinate natural things . . . is possible only by means of mathematics. And since in every doctrine of nature only so much science proper is to be found as there is a priori cognition in it, a doctrine of nature will contain only so much science proper as there is applied mathematics in it."

—METAPHYSICAL FOUNDATIONS OF NATURAL SCIENCE

Preface

It is not necessary to come from the West to be bound inescapably to one's own history. Philosophers, too, beat on, boats against the current, borne back ceaselessly into the past. Inevitably they carry some of the present with them. There is at least this much truth to Croce's dictum that all history is contemporary history.

In this study, I have tried to reconstruct rationally Kant's theory of science, drawing largely on the *Critique of Pure Reason* and downplaying the fact that his position changed and developed over the years. If anything justifies my title, it is the attempt to give a general account of this "theory" and to show how, for Kant, it clarifies the relations between philosophy and science. At the same time, the title "Kant's Theory of Science" is both too broad and too narrow. It is too broad because there are many features of Kant's theory of science, particularly having to do with methodological questions and the role of reason in its regulative aspect, about which I have said very little. There is already an excellent discussion of these questions and this role in Gerd Buchdahl's *Metaphysics and the Philosophy of Science*. The title is too narrow because the differences between Kant's theory of science and his more general metaphysics of experience are ones of detail and emphasis only. What I have to say about the former should illuminate the latter as well.

The reconstruction has been guided by three criteria of adequacy formulated by Professor Wolfgang Stegmüller.[1] He asks that a reconstructed philosophical theory be presented (1) in such a way that it remains in accordance with the basic

[1] In a pair of clarifying papers, "Towards a Rational Reconstruction of Kant's Metaphysics of Experience," *Ratio*, I, 1967, pp. 1-32, and II, 1968, pp. 1-37. Professor Stegmüller has also directed my attention to some little-known but extremely interesting lectures on Kant given by Professor Heinrich Scholz in the winter semester, 1943-1944, at the University of Münster, very much in the same spirit as my study, if not always with the same results.

ideas of the philosopher; (2) as far as possible in precise terms; (3) if possible, as a consistent theory. I would supplement the first criterion, as does Stegmüller, with the proviso that the original text be first analyzed carefully and respected throughout,[2] and extend the third to read "as a consistent and unified theory." There are inconsistencies in Kant's theory; he never denied it. But a problem deeper than exposing or attempting to resolve such inconsistencies is to see how the various themes he develops fit together. In the case of the theory of science, this comes down to trying to develop a unified account of Kant's philosophy of mathematics and his philosophy of physics or, from a different point of view, to relate the *Metaphysical Foundations of Natural Science* to the *Critique of Pure Reason.*

It is also the case that to reconstruct a philosophical position is to make clear the structure of argument that supports it, and this involves as a natural consequence entering into questions of logical theory. A great deal of attention has been devoted to Leibniz's conception of logic, even to the extent of saying, as does Russell, that his metaphysics reflects certain deeply held beliefs about logical form. But comparatively little attention has been devoted to Kant's logical doctrines, apart from routine derogatory remarks about the Table of Judgments in the first *Critique* and the notorious comment that logic had not advanced a step beyond Aristotle.[3] In part, this neglect stems from the fact that Kant, unlike Leibniz, made no distinctive contribution to the history of logic. More important is the fact that some of Kant's most interesting logical doctrines are no more than implicit in his work, and can

[2] I have not hesitated to use contemporary ideas in the attempt to clarify Kant's position; but I have also tried to make clear which of these ideas are in the text and which are introduced from outside in the attempt to cast light on it.

[3] A comment belied by Kant's efforts to break out of the traditional framework, in his essay "On the false subtlety of the four figures of the syllogism," for example, more importantly but less systematically when he finds the framework cramping, as in his discussion of the first two Antimonies in the *Critique of Pure Reason*, where he proposes a distinction between "analytical" and "dialectical" contradictories.

be attributed to him only inferentially. Yet, as I hope to show, they provide important axes and motives of his position.

Understanding Kant's logical theory has been greatly facilitated by a number of recent technical developments. I have tried to make use of these developments, in an informal way, without presupposing any extensive technical background on the part of the reader. Indeed, my own interest in Kant's theory of science was in part prompted by these developments, and the light they shed on the positivist objections to that theory, just as the interest of the neo-Kantians of Marburg was in part prompted by developments in physics around the turn of the century.

Most books on Kant grow out of a more or less sustained attempt to teach the *Critique of Pure Reason*. So did this one. I owe a debt of thanks especially to students in seminars at the University of California, Irvine, and the Claremont Graduate School for their help in what often proved to be a cooperative project.

I owe a debt of thanks as well to former colleagues in the philosophy department at Irvine. In the preface to his book, *Mind and Art*, Guy Sircello has summarized eloquently what it meant to be a member of that department. A present colleague, James W. Allard, has been kind enough to read through several chapters and has made some valuable comments.

A quartet of philosophers have, as teachers and friends, shaped my intellectual development. I would like to thank them while at the same time discharging them of any responsibility for the ways in which their influences are embodied here. It is fair to say that the position of each is, in important respects, Kantian. Donald Davidson provided the general philosophical framework, particularly as concerns its antireductionist impulse. Jaakko Hintikka introduced me to Kant and to the possibilities of exploiting developments in logic to the better understanding of the history of philosophy. Bas van Fraassen supplied some of the developments here exploited. He also read through an early version of the manuscript and made a number of very useful suggestions. Finally, a large portion of this study originated in conversations, over the past several years, with Karel Lambert. Many of his views are reflected here, although, I am afraid, not always very adequately. He has suggested points, patiently corrected mistakes, tried to smooth out the style, and, not least important, provided continuing encouragement. I am very grateful.

Through their system of grants for Younger Humanists, the National Endowment for the Humanities afforded me a very much appreciated year in Munich to work on the manuscript, although the grant is so excellently conceived that it allows for other sorts of work as well. A further grant, also very much appreciated, was provided by Montana State University to bring the manuscript to completion.

It has meant a great deal to me, as it has to other authors of Princeton publications, to have had the continuing interest and encouragement of the Philosophy Editor of Princeton University Press, Sanford Thatcher. I am thankful as well for the care and concern of R. Miriam Brokaw in making the book "objectively real."

Gordon G. Brittan, Jr.
Mission Creek
Livingston, Montana

Contents

KANT'S THEORY OF SCIENCE

Chapter 1: the anti-reductionist Kant

ALMOST everyone follows Hegel in thinking that the history of modern philosophy has a nice symmetry about it. Rationalist thesis ("knowledge is based on reason") gives way to empiricist antithesis ("knowledge is based on sense experience"), which in turn gives way to Kantian synthesis ("knowledge is a product jointly of understanding and sensibility"). For those who play the numbers game, the impression of symmetry is heightened by the fact that there are as many empiricists (Locke, Berkeley, Hume) as rationalists (Descartes, Spinoza, Leibniz), and one Kant, a magical seventh, to reconcile both traditions.

This characterization is oversimplified, of course, but there is nothing essentially wrong with it. The history of modern philosophy, writ large, does have just this sort of dialectic. No one was more aware of the dialectic than Kant himself, as the closing paragraphs of the *Critique of Pure Reason* indicate. It continues to provide a useful framework for the discussion of his ideas. Yet, at the same time, there is another fundamentally important and, it seems to me, largely overlooked way in which what Kant wrote constitutes not so much a reconciliation of "rationalism" and "empiricism" as the rejection of a feature they share in common. This rejection is connected with what I am going to call, somewhat ponderously, the "anti-reductionist" theme in Kant's thought. The present chapter is concerned with developing the theme in some detail. In my view, Kant's theory of science is best approached by first appreciating its anti-reductionist motives.

reductionism characterized

Return for a moment to our rhetorical version of the history of modern philosophy. We begin with Descartes. One of the characteristic claims he wants to make is that the sentences that comprise the various sciences all follow from a few basic propositions, perhaps ultimately the "Cogito," which reason discovers by reflecting on itself. Thus, to retrace briefly a very well-known route, from the fact of his existence Descartes "proves" the existence of God, then, a little further along, that the essence of matter is extension, that momentum is conserved, and eventually Galileo's laws of motion.[1] Similarly, Leibniz claims that, given the principles of contradiction and sufficient reason, all of "natural philosophy" can be "demonstrated."[2] To choose but one example, Leibniz derives Snell's Law ("sin *i*/sin *r* is constant for any pair of media," where *i* is the angle of incidence of a light ray and *r* is its angle of refraction) from the innately given principle of simplicity, a corollary of the principle of sufficient reason, for the law describes the *least path* through a pair of media.[3] The role of sense experience in acquiring scientific knowledge, of which physics is the paradigm, is apparently limited to prompting reason to self-reflection. In other words, the whole of scientific knowledge can be spun out *a priori*, by following up the deductive implications of those basic principles which reason originally discloses to itself.[4]

[1] Descartes writes, for example, in the *Principles of Philosophy*, part IV, #203 (French version only), that all of physics follows from the self-evident (clear and distinct) proposition that matter is extension. In fact, it is not entirely clear whether Descartes thought that physical laws could be derived *a priori*. He suggests in the preface to the *Principles* that experiments are necessary to "support and justify" his reasoning, although these words could be understood in a sense compatible with the claims made at IV, #203, and elsewhere in the preface (e.g., from the basic principles of philosophy "may be derived a knowledge of all things that are in the world").

[2] See the first paragraph of his second letter to Clarke, in *The Leibniz-Clarke Correspondence*, ed. H. G. Alexander (Manchester: Manchester University Press, 1956).

[3] See his "Tentamen Anaqoqicum: An Anagogical Essay in the Investigation of Causes," in *Leibniz: Philosophical Papers and Letters*, ed. L. Loemker (Dordrecht: D. Reidel Publishing Company, 1969).

[4] Leibniz's follower Christian Wolff, for Kant "the greatest among all dogmatic philosophers," attempted not only to found physics on the principle of sufficient reason, but to prove *that* principle in turn from the principle of contradiction. See his *Ontology* (1736), #70. Others in the same tradition include Schelling who, according to C. I. Lewis, *Mind and the World Order* (New York: Dover Publications, Inc., 1956), p. 190n., "starting from the Fichtean premise, A = A, . . . deduces eventually the electrical and magnetic properties of matter."

This is only a very rough sketch, but it indicates two important aspects of the view I am attributing to Descartes and Leibniz. First, that physics in its entirety can be derived from a few basic logical and metaphysical principles and, second, that these principles are discovered and their truth guaranteed by reason. If the second aspect of their position is primarily what makes Descartes and Leibniz "rationalists" (in perhaps a not quite standard sense of the word), the first makes them—to coin a term—"reductionists." Provisionally, a "reductionist" is someone who claims that a deductive relationship obtains between sentences of one kind and sentences of another, "kinds" to be made clearer by way of examples. Descartes' attempted derivation of physical laws from metaphysical principles is an attempted reduction of the former to the latter. So too would be the derivation of biology from physics that he forecasts. In the same way, mathematics reduces to logic if the propositions of mathematics can be derived from the laws of logic and definitions, as Leibniz believed.[5]

Classical empiricists do not, to my knowledge, claim that a deductive relationship obtains between the propositions of physics and sentences reporting or describing certain sorts of primary sense experience.[6] This is not to say that there are no texts that at least superficially claim an empiricist reduction. Among the most striking of these is the general Scholium added to the second edition of *Principia*, where Newton declares: "I frame no hypotheses; for *whatever is not deduced from the phenomena* is to be called an hypothesis; and hypotheses, whether metaphysical or physical, whether of occult qualities or mechanical, have no place in experimental philosophy. In

[5] Attempted derivations of this kind are familiar to students of contemporary philosophy, much of whose history revolves around various reductionist programs—phenomenalism, behaviorism, logicism, operationalism, and so on. So central have these programs been, in fact, that "philosophical analysis" has, among many of its practitioners, come to be virtually identified with providing a reduction of the types indicated.

[6] Although it is certainly suggested by Berkeley and Hume. Their views will be discussed in more detail in chapter 5. Mill was perhaps the first to advance an explicit phenomenalism.

this philosophy particular propositions are *inferred* from the phenomena. . . ."[7] There is also the equally celebrated Query XXXI at the end of the second edition of Newton's *Optics*: "To tell us that every species of things is endowed with an occult specific quality by which it acts and produces manifest effects is to tell us nothing; but *to derive two or three general principles of motion from phenomena*, and afterwards to tell us how the properties and activities of all empirical things follow from these manifest principles, would be a very great step in philosophy. . . ." The difficulty with such texts is that the significance of the words "deduced," "derived," and "inferred," even "phenomena," is far from clear, however we might broaden our characterization of reductionism.[8]

But it is worth adding that if classical empiricists are not explicit about claiming that a deductive relationship obtains between the propositions of physics and sense-experience sentences, many of their more recent successors have insisted on it. Ernst Mach for one, C. I. Lewis for another, Rudolf Carnap for a third[9] have claimed that the propositions of physics can be "translated" or "transformed" into sense-experience sentences, and, while the notion of translation involved here is not always clear, at the very least it must mean that sentences, or sets of sentences, of the two kinds are deductively equivalent. Once again, to the extent that the propositions of physics can be derived from sense-experience sentences, physics has been reduced to them.

As I have formulated them, "rationalism" and "empiri-

[7] *Mathematical Principles of Natural Philosophy and His System of the World*, the Motte translation (1729), revised by Florian Cajori (Berkeley: University of California Press, 1962). My italics.

[8] Although some historians of science are not thereby deterred from putting Newton in the middle of the positivist tradition. E.g., Mary Hesse, *Forces and Fields* (New York: Philosophical Library, 1962), p. 2: "he wished to confine theories to what could be *deduced* from the phenomena." Interestingly, Professor Hesse puts Kant in the same tradition. My italics.

[9] In *The Science of Mechanics* (LaSalle: Open Court Publishing Company, 1902), chapter IV, section IV, *An Analysis of Knowledge and Valuation* (LaSalle: Open Court Publishing Company, 1946), chapter VIII, and *The Logical Structure of the World* (Berkeley: University of California Press, 1967), respectively.

cism" share a reductionist feature, although they differ sharply concerning the character of the sentences to which the reduction is to be made. They also share a common motive. In both cases a reduction is typically carried out with the purpose of providing physics with a firm epistemological foundation. If physics can be reduced to ideas innate to reason, whose truth and necessity is thereby guaranteed, then physics will have been shown to be epistemologically secure. For those of empiricist persuasion, the demand is that physics be reduced to sense experiences that are themselves incorrigible; only in this way can the knowledge claims that science makes be justified. Or, if such a reduction cannot be carried out, we should in conscience become skeptics.

reductionism rejected: "the Copernican Revolution"

It is just at this point that I want to locate Kant's break with the philosophical tradition. The break has several different aspects, all of which can be brought together under the very general heading of "the Copernican Revolution" Kant hoped to bring about in philosophy. For the moment, it is not so much a question of giving arguments as of sketching a changed perspective.

In the first place, the reductionist program, and even more strikingly its epistemological motive, is abandoned. Kant begins with the security of physics. It is the point of departure, not the end of a long demonstration. There is no need to justify the propositions of physics, individually or collectively, or to reduce them to something more basic or secure. We are not to ask whether we do in fact have genuine knowledge of nature, but rather, granting from the outset that we do have such knowledge, what conceptual abilities are *presupposed*. Or, as Kant himself preferred to put it, not "Is knowledge possible?" but "How is it possible?"

I have suggested that for both empiricists and rationalists the epistemological security of physics typically motivated reductionist programs. Why the worry? For one thing, there were traditional problems about the reliability of sense perception, most of them pointed up by one version or another of the "argument from illusion." For another thing, and for my purposes the more important, two fundamental theses of 17th- and 18th-century science were taken to imply that our knowledge of the world could never be more than inferential. The first of these theses had to do with a physiological account of perception. On this account, physical objects impinge on the various sense organs, giving rise to sensations that are in turn the immediate objects of perception. The second thesis was that, in reality, the world is composed of imperceptible particles or corpuscles. The two theses were combined in the claim that these imperceptible particles had merely spatial properties, and that the color, taste, and smell we ordinarily attribute to the objects in our environment are in some sense subjective, dependent on peculiarities of human physiology. Both theses apparently open up a gap between the world as it presents itself to our senses and untutored intelligence, and the world as it actually is, in the light of scientific investigation. If physics is taken to describe the world as it actually is, then it would appear, possibly with some additional assumptions, that its epistemological security can be guaranteed only if this gap is bridged or closed. A central aim of reductionist programs, of both empiricist and rationalist varieties, is to provide the appropriate closure or bridge. Talk about the world as it emerges in the scientific picture of things, and in particular talk about imperceptible particles, is legitimized only insofar as it can be translated into talk about sense experiences, on the one hand, or shown to follow from necessary first principles, on the other.

Kant subscribed to the theses of 17th- and 18th-century science just mentioned,[10] but he denied that an epistemological gap opened as a result, a gap that philosophers had to close or bridge. We can begin with the second thesis, that the world is composed of imperceptible particles. In a passage at B226/B273 of the *Critique of Pure Reason* the following point is made: "Thus from the perception of the attracted iron filings we know of the existence of a magnetic matter pervading all bodies, although the constitution of our organs cuts us off

[10] Construing the word "particle" *very* broadly, for Kant was not an atomist. See chapter 6.

from all immediate perception of this medium (*dieses Stoffs*). For in accordance with the laws of sensibility and the context of our perceptions, we should, were our senses more refined, come also in an experience upon the immediate empirical intuition of it. The grossness of our senses does not in any way decide the form of possible experience in general."[11]

The point is complex, but it seems to have at least two aspects. One is the explicit denial that the imperceptibility of the medium (the magnetic matter) creates an epistemological gap. The other is an intended contrast, made clearer in the paragraphs following the passage quoted, between this type of inference, made within the limits of possible experience, and a pretended inference from what is given in experience—sensations, for example—to what is beyond experience, those objects lying on the other side of the veil of perception. The latter sort of inference, Berkeley rightly saw, can never be justified, not even by an appeal to God's benevolence. We must distinguish the former, legitimate (and properly scientific) inference from the latter, illegitimate (and metaphysical) one, the contingent imperceptibility that depends on the grossness of our senses from the non-contingent imperceptibility that has to do with the limits of possible experience. If we do not make the distinction, and we do take seriously the scientific picture of things, then there would seem to be no option to idealism, "the theory which declares the existence of objects in space outside us either to be merely doubtful and indemonstrable or to be false and impossible."[12] On the other hand, Kant goes on to argue, in the Refutation of Idealism that immediately follows the quoted passage in the *Critique*, not only is the existence of objects in space outside us in general not doubtful, hence no inference and, *a fortiori*, no "reduction" is needed, it is in some sense more certain than awareness of private sense experiences or the fact of one's own existence. The motives that prompt reductionist accounts inevitably

[11] This and all subsequent quotations of the *Critique of Pure Reason* are from the translation by Norman Kemp Smith (London: Macmillan and Co., Ltd., 1933). I follow the usual practice of referring to the first and second editions of the *Critique* as "A" and "B" respectively.

[12] *CPR*, B274.

lead to idealism. Kant's rejection of the one is of a piece with his rejection of the other.[13]

In the same way, Kant denies that an epistemological gap between ourselves and the world opens as a result of the first thesis, that perception is causally mediated. As he puts it in the General Observations on Transcendental Aesthetic, at A45/B62: "We commonly distinguish in appearances that which is essentially inherent in their intuition and holds for sense in all human beings, from that which belongs to their intuition accidentally only, and is valid not in relation to sensibility in general but only in relation to a particular standpoint or to a peculiarity of structure in this or that sense. The former kind of knowledge is then declared to represent the object in itself, the latter its appearance only."[14] But, Kant adds at once, this distinction is merely empirical. It is a distinction made within experience, not between experience and that which lies beyond or behind it. It is, in fact, only if we assume the existence of physical objects impinging on our

[13] One might object that my emphasis on Kant's empirical realism ignores his "transcendental idealism" and therefore is misleading, if not also problematic, with respect to his break with his philosophical predecessors and to his reconstruction of physics. To discuss "transcendental idealism" in any detail would take us rather far afield from the questions I want to discuss; in fact, I do not see how focussing on the "idealist" themes in Kant's thought (e.g., the appearance/thing in itself contrast) would alter any of the conclusions reached. But the following familiar points might be noted: (a) Kant is very careful to distinguish his "idealism" from that of, e.g., Descartes and Berkeley (see note III to #13 of the *Prolegomena*); (b) he insists that transcendental idealism *entails* empirical realism (*Critique of Pure Reason*, A371); (c) his "idealism" properly concerns the status of space and time; (d) his point is that unless we assume that the spatial-temporal structure of the world is in some sense contributed by us (it is not an object but constitutive of the objects we can experience), there is no way to guarantee the fit of mathematics to the world, hence no way to secure the objective validity of physics; (e) Kant's view does not commit him to two sorts of objects, (unknowable) things in themselves and (knowable) appearances, but only to two ways of looking at objects, in relation to the kind of knowledge we can have about them, as the note to Bxvii in the *Critique* and the chapter on Phenomena and Noumena make clear.

[14] "Object in itself" and "appearance" are not used here in Kant's technical way, but have their more ordinary meanings.

sense organs from the outset that a physiological account of perception is coherent.[15]

With regard to both theses, Kant often makes his point in terms of a distinction between transcendental and empirical questions. Kant uses the word "transcendental" in several different ways, often carelessly. On one of these uses, the distinction between transcendental and empirical questions is a distinction between questions *about* experience and questions asked *within* experience, a distinction, we might put it today, between the kinds of questions philosophers ask and the kinds of questions scientists ask. Both the physiological theory of perception and the corpuscular hypothesis are empirical or scientific hypotheses, and the distinction they suggest between the world as it is and the world as it appears is empirical. It is when this distinction is taken as a transcendental distinction, between what lies within our experience and what beyond it, that an epistemological gap is created, attempts at reduction initiated, and idealism produced. Part of Kant's break with his predecessors (and, I think, with certain aspects of his own pre-Critical writings on science) involves his claim that they confused empirical with transcendental questions.

Kant's claim that the propositions of physics stand in no need of justification is connected in another way with his view of what it is to do theory of science or, more generally, of what it is to do philosophy. On the traditional view, to do theory of science is to *re-do* science, to reconstruct it rationally so that it is clear how its content may be guaranteed against skeptical attack. This is also the view of many of the more recent logical positivists: philosophy consists in the analysis of the propositions of science, and to analyze a scientific proposition is to show how it can be "reduced" to immediate sense experience. But Kant rejects this view of what it is to do theory of science. In his view, theory of science is not concerned so much with the justification of phycics—indeed, physics stands in no need of justification—as with the isolation and examination of the *a priori* and conceptual elements

[15] Graham Bird, *Kant's Theory of Knowledge* (London: Routledge & Kegan Paul, 1962), chapter 2, discusses this sort of point persuasively and at length.

in it and the role they play in making physics "possible," that is, in guaranteeing its application to objects that in an important sense are independent of us. Not the epistemological security of physics but its "objectivity" is what is at stake.[16] This changed conception is still another aspect of "the Copernican Revolution."

These claims about Kant's position are, of course, very general and so far largely unsupported. They are intended merely as a suggestive preamble to what follows. Even so, they run counter to two standard ways in which Kant's position has been interpreted. One of these interpretations bases itself on passages in the *Critique of Pure Reason*, where a kind of phenomenalism is suggested,[17] and on a particular reading of the doctrine of "synthesis," a reading associated with the picture of the mind "putting together" sensations to "form" objects. This interpretation derives some additional support from W. V. Quine's well-known thesis that classical empiricism turns on two inseparable dogmas: the analytic/synthetic distinction and empiricist reductionism. Kant insists on the analytic/synthetic distinction. If Quine is right, it would follow naturally (if not logically) that he is committed to a reductionist view of the empiricist variety.[18] The other interpretation mentioned again bases itself on passages in the *Critique*, this time where the mind is pictured as laying down the law to nature,[19] and on a particular reconstruction of

[16] "Objectivity," for Kant, requires objects, not a certain kind of evidence. The contrast between epistemological security and "objectivity" will be explored further in chapter 5.

[17] For example, at A110-114 ("But when we consider that nature is not a thing in itself but is merely an aggregate of appearances, so many representations of the mind . . .").

[18] As a long line of commentators, although with important variations between them, have maintained. A relatively recent example, Jonathan Bennett, *Kant's Analytic* (Cambridge: Cambridge University Press, 1966), p. 22: "Kant thinks that statements about phenomena are not merely supported by, but are equivalent to, statements about actual and possible sensory states."

[19] For example, at A125: "Thus the order and regularity in the appearances, which we entitle *nature*, we ourselves introduce. We could never find them in appearances, had we not ourselves, or the nature of our mind, originally set them there. For this unity has to be a necessary one, that is, has to be

Kant's central argument. It is to the effect that Kant is a "reductionist" in the rationalist tradition, attempting to derive, *inter alia*, the whole of Newtonian physics from certain necessary propositions about human sensibility and understanding.[20]

There are a variety of things wrong with both of these interpretations. For the moment, however, I want to concentrate on two of their respective sources of support, Quine's thesis and the rationalist reconstruction of Kant's argument. Consideration of the first should help us to become clearer about Kant's characterization of the analytic/synthetic distinction and the attendant concept of synthetic *a priori* judgments. It will also allow us to raise questions concerning the revisability of such judgments. Consideration of the structure of Kant's argument, on the other hand, should help us to become clearer about his motives and about the key relation of presupposition. Between them, the analytic/synthetic distinction, the concept of synthetic *a priori* judgments, and the presupposition relation provide most of the scaffolding of Kant's Critical position. The discussion in succeeding chapters is in large part application of my remarks about them here.

the analytic/synthetic distinction and really possible worlds

It is worth quoting Quine's presentation of his thesis at some length, particularly since I am going to adapt it to my own ends:

"The dogma of reductionism . . . is intimately connected

---

an *a priori* certain unity of the connection of appearances; and such synthetic unity could not be established *a priori* if there were not subjective grounds of such unity contained *a priori* in the original cognitive powers of our mind, and if these subjective conditions, inasmuch as they are the grounds of the possibility of knowing any object whatsoever in experience, were not at the same time objectively valid."

[20] As another long line of commentators, again with important variations, have maintained. A relatively recent example, Sir Karl Popper, *Conjectures and Refutations* (New York: Basic Books, 1962), p. 180n.: ". . . some of the greatest difficulties in Kant are due to the tacit assumption that Newtonian science is demonstrably true. . . ."

with the other dogma—that there is a cleavage between the analytic and the synthetic. We have found ourselves led, indeed, from the latter problem to the former through the verification theory of meaning. More directly, the one dogma clearly supports the other in this way: as long as it is taken to be significant in general to speak of the confirmation and infirmation of a statement, it seems significant to speak also of a limiting kind of statement which is vacuously confirmed, *ipso facto*, come what may; and such a statement is analytic.

"The two dogmas are, indeed, at root identical. We lately reflected that in general the truth of statements does obviously depend upon both language and upon extralinguistic fact, and we noted that this obvious circumstance carries in its train, not logically but all too naturally, a feeling that the truth of a statement is somehow analyzable into a linguistic component and a factual component. The factual component must, if we are empiricists, boil down to a range of confirmatory experiences. In the extreme case where the linguistic component is all that matters, a true statement is analytic."[21]

The basic idea seems to be that those who distinguish between analytic and synthetic sentences do so on the basis of the criterion that synthetic but not analytic sentences have a sense experience translation.[22] But to say that synthetic sentences have a sense experience translation is to subscribe to the dogma of reductionism, although not necessarily to the motives I gave for it earlier. We can try to rephrase this,[23] with a view to making the connection more precise. The "dogma of reductionism" is that from consistent and possibly infinite sets of sentences describing or reporting sense experiences we can derive the class of synthetic sentences, in particular the propositions of physics. But no two synthetic sentences thus "reduce" to the same set of sense-experience sentences; to say that there are two such propositions is to say that different

[21] *From a Logical Point of View* (Cambridge: Harvard University Press, 1953), p. 41.

[22] Quine puts the point in terms of ranges of confirmatory and disconfirmatory experiences, but I think the point is essentially the same.

[23] Again to use my terms, not Quine's, but hopefully not at the expense of distorting his position.

sense experiences are relevant to the truth of each. On the other hand, from *any* set of sense experience sentences we can derive *any* (not self-contradictory) analytic sentence. To characterize the analytic/synthetic distinction in this way is thus, if one also holds that there are in fact sentences of each kind, to subscribe to the dogma of reductionism. One cannot make the distinction on this basis and fail to be a reductionist.

The distinction Kant makes between analytic and synthetic sentences, and on the importance of which he continually insists,[24] is not set out in just this way. It is notoriously difficult, in fact, to say just how the distinction he makes is to be characterized, independent of the metaphors to which he so often retreats. Kant initially characterizes analytic sentences in the *Critique of Pure Reason* as "those in which the connection of the predicate with the subject is thought through identity" (A7/B10). In the same paragraph, he says that in such sentences the predicate is "contained" in the subject. The "containment" metaphor used here seems to pick up on the "analytic" character of the sentences in question by way of a chemical model; what we do is to "decompose" the subject term, submit it to analysis, to see what it contains. The sentences to which it most easily lends itself are those like "Mad dogs are dogs," where in an obvious way the predicate is "contained" in the subject, and those like "All bachelors are unmarried" where the subject can be "decomposed" into constituents (in this case "unmarried man") one of which is the predicate. On the other hand, the "containment" metaphor appears to be limited to sentences of subject predicate form,[25] and it appears

[24] As he says in the note to #3 of the *Prolegomena*: "This division is indispensable in respect of the critique of human understanding, and hence deserves to be classical in it. . . . " This and all subsequent quotations of the *Prolegomena to Any Future Metaphysics* are from the Peter Lucas translation (Manchester: Manchester University Press, 1953). Future page references to the *Prolegomena* are to the Prussian Academy edition, indicated as well in the Lucas edition.

I am using the word "sentence" interchangeably with "statement" and "proposition" instead of Kant's word "judgment" to indicate the object of judgment rather than the act of judging.

[25] Hence does not readily accommodate such clearly analytic sentences as those of the form "*p* or not *p*."

to be applicable only when general (i.e., complex) concepts stand as subject terms. Yet at B16, Kant says that any judgment of the form "*a* = *a*" is analytic, hence, for example, "Kant = Kant" is. But in what sense is "Kant" analyzable and how is "is identical to Kant" contained in it? Kant, the person not the name, gives us no clues.[26]

The "containment" characterization in any case gives way to another: analytic judgments are those the negations of which are self-contradictory.[27] We might put this by saying that all analytic judgments are "reducible" to the principle of contradiction. Since this is precisely the characterization Kant gives of the truths of logic as well,[28] I think we are justified in venturing the following reconstruction of the analytic/synthetic distinction. Those sentences are analytic which are logical truths or which can be turned into logical truths upon the substitution of synonyms for the terms they contain. Thus, "John is tall or it is not the case that John is tall" is a logical truth (it is an instance of the valid schema "*p* or not *p*"), hence it is analytic. "All oculists are eye-doctors" is similarly analytic, for it can be turned into the logical truth "All oculists are oculists" (an instance of the valid schema "$(x)\ (Fx \rightarrow Fx)$") on substitution of the term "oculist" for its synonym "eye-doctor." And the same thing can be said about Kant's paradigm sentence "All bodies are extended."

Kant characterizes logical truths, we have just seen, in terms of the principle of contradiction. I want now to suggest that it throws a great deal of light on Kant's position to characterize logical truths in terms of a second notion, that of

[26] The analyticity of "a = a" poses problems to which we shall return in the next chapter.

[27] "The *principle of contradiction* must . . . be recognized as being the universal and completely sufficient *principle of all analytic knowledge*" (A151/B191).

[28] See his lectures on logic (first given in 1765, but continuously revised over the years and eventually published by Kant's student Jäsche in 1800), for example under the heading "Criteria of Logical Truth," Academy edition, IX, pp. 70ff. An English translation of the *Logic*, by R. S. Hartman and W. Schwarz (New York: The Bobbs-Merrill Company, Inc., 1974) has recently appeared.

truth in all possible worlds. This notion does not appear in Kant,[29] and, as we shall see in more detail in the next section of this chapter, there are reasons why he might balk at introducing any talk of "worlds" into logical theory.[30] But, used informally and heuristically, it will, I think, eventually justify its introduction.[31] Since the characterization of logical truth in terms of the notion of truth in all possible worlds is crucial to my reconstruction of Kant's analytic/synthetic distinction, I want to try to clarify it somewhat before going on.

Clarification involves taking a closer look at the position of Leibniz, who first proposed the characterization. Leibniz's most complete discussion of the modalities, and of the attendant notion of a possible world, seems to come in the *Theodicy*, although, as is usual with Leibniz, much of what is most interesting in his view occurs in fragments and letters. In Leibniz's view, very briefly, the actual world is only one of an innumerable number of worlds that could have existed; each one of these worlds is, in this sense, a possible world. To say that a world is possible is at the same time to say that its description is self-consistent (or "conceivable"), for as Leibniz puts it in #173 of the *Theodicy*, everything that implies a contradiction is impossible, and everything that does not imply a contradiction is possible.

Leibniz illustrates his view in the *Theodicy*, ##413ff., by way of a myth. A certain Sextus complains of his fate to Zeus, who in turn sends him to Athens to see the goddess Pallas Athena. She shows him through a palace in which there are infinitely many rooms, each room constituting a cosmos or possible world. In one room, Sextus sees a world displayed in

[29] Except in his first published essay, *Thoughts on the True Estimation of Living Forces* (1747), where he makes use of it, insisting on the existence of an innumerable number of possible worlds (#8) and adding, against Leibniz, that these worlds are characterized by spaces of varying dimensions.

[30] Since a *world* is an "Idea of Reason," beyond the bounds of sense.

[31] Robert Howell, "Intuition, Synthesis, and Individuation in the *Critique of Pure Reason*," *Nous*, VII, pp. 207-232, and Philip Kitcher, "Kant and the Foundations of Mathematics," *Philosophical Review*, 84 (1975), pp. 23-50, both make use of the concept of a "possible world" in reconstructing Kant's position, although along somewhat different lines.

which he goes to Corinth, finds a treasure, and lives a life of wealth; in another room, he sees a world displayed in which he goes to Thrace, marries the king's daughter, and succeeds to the throne; and so on. Only one world is actual, the world in which Sextus in fact lives, but all are possible insofar as they contain consistent Sextus-biographies.

There is one more point about Leibniz's view that should be mentioned. There is room in the text to interpret a possible world either as a set of objects together with a set of properties of these objects or as a set of (singular) concepts together with a set of attributes. Just as Leibniz seems to vacillate between these "object" and "concept" characterizations of possible worlds (although the second seems more compatible with other aspects of his position)[32] so too will I. No serious confusion should result.

Since, following Leibniz, a possible world is a world whose description is self-consistent, there is a close connection between the characterization of logical truths as sentences true in all possible worlds and that given earlier, in terms of the principle of contradiction. It is thus possible to read both characterizations back into the text, for example, in this passage at A151/B190 of the *Critique*: "The proposition that no predicate contrary of a thing can belong to it, is entitled the principle of contradiction, and is a universal, though merely negative, criterion of all truth. For this reason, it belongs only to logic. It holds of knowledge, merely as knowledge in general, irrespective of content." For to say that a sentence holds "irrespective of content" is just to say that it holds in or of all possible worlds, without respect to any further differences between them. The same kind of characterization of logical truth, in terms of content and objects, is repeated in a number of different places. For example, at A52/B76: "logic treats of understanding without any regard to differences in the objects to which the understanding may be directed." And at A152/B191: ". . . this famous principle (i.e., of contradiction) is

[32] See Benson Mates's illuminating paper, "Leibniz on Possible Worlds," in B. van Rootselaar and J. F. Staal, eds., *Logic, Methodology, and Philosophy of Science*, IV (North-Holland Publishing Company, 1968), pp. 507-529.

thus without content and merely formal. . . ." Finally, in the opening paragraphs of his lectures on logic, Kant heavily emphasizes the "merely formal" and "contentless" and "universally applicable" character of logical truths.[33] But at the same time, since analytic sentences are those which cannot be denied consistently, it follows that they too must be true in or of all possible worlds.[34] What is important about the "containment" metaphor in the final analysis is, first, that whether a predicate is contained in the subject is a formal matter, and can be determined by inspection (on analysis), and, second, that since the predicate is already contained in the subject, analytic judgments are "without content," providing us with no new information. By contrast, purely synthetic judgments are true in or of the actual world, and perhaps some other possible worlds, but they are not true in or of all possible worlds generally. To the extent that they do not hold "irrespective of all difference in objects" they have content and are not merely formal and "empty."

I have already indicated that, prior to its reconstruction, the

[33] Logic is characterized in the introduction to the lectures on logic as "eine Wissenschaft a priori von den notwendigen Gesetzen des Denkens, aber nicht in Ansehung besonderer Gegenstände, sondern aller Gegenstände überhaupt," Academy edition, IX, p. 10.

[34] Kitcher, "Kant and the Foundations of Mathematics," p. 24, suggests the reason: "What features a world may have are limited by the structure of our concepts. Some propositions are true in each world in virtue of this limitation. In a derivative sense, these propositions can be said to be true in virtue of the structure of our concepts because they owe their truth to particular features of that structure. Kant calls these truths 'analytic.' " In much of the recent, particularly Anglo-Saxon, literature on Kant it is suggested that for Kant analytic sentences are those "true in virtue of the meanings of the words they contain." As it stands, and despite frequent repetition, this is a vague phrase. Kant does say that analytic judgments are "valid from concepts," but this characterization is intended to draw our attention to the fact that they are *explicative*, not *ampliative*, and does not commit him to the truth-in-virtue-of-meanings formula. Indeed, while some of Kant's examples ("All bodies are extended") invite the application of the formula, others, notably the principle of contradiction itself, definitely exclude it. Kant certainly did not think that principle was true in virtue of the meanings of "and" and "not," although it does owe its truth in some sense to the way in which our concepts structure the class of possible worlds.

characterization Kant actually gives of the analytic/synthetic distinction, especially in the opening pages of the first *Critique*, is somewhat unclear. It should be added at once that part of the difficulty lies in the fact that in calling statements analytic or synthetic he intends, at different times, to draw our attention to different aspects of them; his usage is not perfectly consistent. One can even go so far as to say that once having insisted on its importance, Kant only intermittently keeps the distinction, in any of its versions, in mind.[35] But of much greater issue is the fact that for Kant not every proposition is analytic or synthetic *tout court*. Some propositions, apparently including certain propositions of physics, are synthetic *a priori*. "I need only cite two such judgments," he says in the Introduction to the *Critique*, "that in all changes of the material world the quantity of matter remains unchanged, and that in all communication of motion, action and reaction must always be equal." The difficulty is to characterize such propositions in such a way that they are clearly neither *a posteriori* nor analytic and so that the analytic/synthetic distinction is itself not blurred. To provide the characterization has proved to be almost as difficult, in fact, as demonstrating that these propositions play the role with respect to experience that Kant assigns them.

The characterization of synthetic *a priori* judgments I am going to suggest, again "from the outside," requires the introduction of a new and, for Kant's enterprise, fundamentally important concept, that of "real possibility." There are a variety of passages in which this concept is discussed;[36] many of them will be referred to later. Among the more explicit is the note at A596/B624, in the section of the *Critique* on the impossibility of an ontological proof of God's existence:

"A concept is always possible if it is not self-contradictory. This is the logical criterion of possibility, and by it the object of the concept is distinguishable from the *nihil negativum*. But it may none the less be an empty concept, unless the objective

[35] See Norman Kemp Smith, *A Commentary to Kant's "Critique of Pure Reason,"* second edition (New York: Humanities Press, 1962), pp. 33ff.

[36] E.g., A75/B101, A220/B267, A239/B298, A244/B302, A290/B347.

reality of the synthesis through which the concept is generated has been specifically proved; and such proof, as we have shown above, rests on principles of possible experience, and not on the principle of analysis (the principle of contradiction). This is a warning against arguing directly from the logical possibility of concepts to the real possibility of things."

This passage has several different features and introduces another concept, that of objective reality, that will prove to be crucial in the sequel. What interests us here, however, is the distinction it sets out between logical and real possibility. Logical possibility, the wider notion, is characterized in terms of the law of contradiction, real possibility in terms of the principles of possible experience, *viz.* the Categories.

The correlate of the concept of real possibility is the concept of a really possible world. Thus we can express the relationship between the concepts of real and logical possibility by saying that while for Kant every really possible world is logically possible, it is not the case that every logically possible world is really possible.

What is a "really possible world"? Provisionally, such a world can be characterized from two points of view. From one, it is a world that beings like ourselves, endowed with certain perceptual capacities and conceptual abilities, could experience. Kant sometimes makes it look like a quasi-psychological *brute fact* that we have these capacities and abilities and that our inherent mental structure somehow determines the form of the worlds we can experience, i.e., really possible worlds.[37] In fact, Kant has an *argument*, the *Transcendental Deduction*, to show that only a certain kind of world is a world we could *experience*. From the other point of view, a really possible world is a world whose limits and general form are given by the Categories, a world having a particular spatial-temporal-causal form that contains enduring centers of attractive and repulsive forces.

Recall that on the present reconstruction of Kant's position,

[37] Immortalized in the notorious (and non-Kantian) image of the man wearing irremovable space-time spectacles, whose perceptual experience as a result is all of a spatial-temporal character.

those sentences are analytic which are logical truths or are transformable into logical truths on substitution of synonyms, and, further, that logical truths are sentences true in or of all possible worlds. Synthetic *a priori* sentences, on the other hand, are sentences true not in all possible worlds, but in all *really possible* worlds. There is no world *we* could experience in which they would not hold; in this sense they are both universal and necessary. There are, for example, possible worlds in which conservation of matter and what looks like Newton's third law of motion do not hold: "in all changes of the material world the quantity of matter does *not* remain unchanged" and "in all communication of motion, action and reaction are *not* necessarily equal" are self-consistent propositions. But there is no *really* possible world in which these latter propositions might hold. Worlds that they described would not be worlds, according to Kant, that we were capable of experiencing.[38]

I have just suggested that synthetic *a priori* truths describe the class of "really possible" worlds. One point Kant wants to make is that a great number of worlds are compatible with the actual world, although they differ from it in one respect or another. The general character of experience that Kant attempts to delineate in the *Critique of Pure Reason* is to be distinguished from the specific character of our actual experience, a contrast he also makes in terms of the form/matter and *a priori/a posteriori* distinctions. As the point is made in the *Inaugural Dissertation*: "the bond constituting the essential form of a world is regarded as the principle of possible interactions of the substances constituting the world. For actual interac-

[38] This theme seems to be present in Kant's thought as early as his essay, *The Only Possible Foundation for a Proof of the Existence of God* (1763). It can be illustrated by Leibniz's mythic palace of Pallas Athena. All the innumerable rooms of the palace are possible worlds; their description is self-consistent. Only some of the rooms are really possible worlds; that is, *we* are capable of experiencing the contents of only some of the rooms, the others being in some sense barred to us (although the word "room" already suggests a possible object of experience). Among the really possible worlds only one is actual; this is the room in which Sextus' true-life adventure takes place.

tions do not belong to essence but to state."[39] Synthetic *a priori* truths define the class of possible (here, as often elsewhere, read "really possible") interactions. To find out which interactions from the class of possibilities are actual, experience is necessary. One cannot otherwise draw a conclusion about what is actual from what is really possible. In fact, the propositions to which Christian Wolff, for example, thought the propositions of physics could be "reduced" were not taken by him to be synthetic *a priori* but analytic, that is, principles derived from the law of contradiction. But these principles characterize what is merely possible, any consistently describable world including our own. There is a great gap between what is possible in this sense and what is actual. We might describe the situation as follows: logic is the physics of possible objects, what Kant calls "transcendental logic" (the system of the Categories) the physics of really possible objects, and physics the physics of physical objects.

Thus, *one* of the things Kant means when he says that physical propositions are "only special determinations of the pure laws of the understanding" is not that the former can be derived from the latter, but that they must be compatible. The actual world is both merely and really possible; in this sense, the laws that describe it are only "special determinations of the laws of the understanding."[40]

Two matters must be kept separate, despite the fact that Kant seemingly runs them together. One is his attempted proof that only a certain kind of world, a "sensible world,"[41] is a really possible object for us. The other is the distinction, fundamental to his project, between possibility and real pos-

[39] From the translation by J. Handyside in *Kant's Inaugural Dissertation and Early Writings on Space* (LaSalle: Open Court Publishing Company, 1929), p. 40.

[40] Both Scholz and Stegmüller, in the works mentioned in the Preface, emphasize this aspect of Kant's position, construing the Categories as limiting the range of possible physical theories between which empirical investigation then decides. On their reconstruction, the Categories serve as rules of eliminative induction.

[41] To borrow from the title of Kant's Inaugural Dissertation, *On the Form and Principles of the Sensible and Intelligible World*.

sibility. We might reject the proof, and the claim that inevitably any object of experience for us must have certain general features, and yet still want to make out a contrast between possible and really possible worlds, the latter being, for example, that set of worlds compatible with the most general principles of physical theory (at a given point in time). Unless we can make such a contrast, in any case, the distinction between analytic and synthetic *a priori* sentences, on which so much of Kant's case depends, collapses.

To summarize, I have suggested that we can distinguish between analytic, synthetic, and synthetic *a priori* sentences in terms of the numbers and types of the "worlds" in or of which they hold. A sentence is analytic if it is true in or of all possible, i.e., consistently describable, worlds.[42] A sentence is synthetic just in case it is not analytic. A sentence is synthetic *a posteriori* true in case it is true in or of the actual world (and perhaps in or of some other really possible worlds as well) but not in or of all really possible worlds; it is in this sense an accidental or contingent truth. A sentence is synthetic *a posteriori* false just in case it is true in or of some really possible worlds and false in the real world. A sentence is synthetic *a priori* true just in case it is true in or of all really possible worlds, including the real world, but it is not true in or of all merely possible worlds, for example, the propositions of Euclidean geometry. A sentence is synthetic *a priori* false just in case it is false in or of every really possible world, for example, the propositions of non-Euclidean geometry.[43]

Now back to Quine. The first point is simply that Kant avoids the "reductionist" tag to the extent that he insists that at least some synthetic sentences, those that are *a priori* as well, do not (in Quine's words) "boil down to a range of

[42] For reasons advanced in the next chapter, we will eventually modify this to read that analytic sentences, if true, are true in or of. . . .

[43] Kant uses the words "*a priori*" and "*a posteriori*" in an epistemological way as well, to indicate *how we come to know* the propositions of the various types. Thus synthetic *a priori* truths are those we can come to know independently of any particular sequences of experiences, although they cannot be known simply by analyzing them into their component concepts. On my interpretation, this aspect of Kant's characterization is not crucial.

confirmatory experiences." These sentences, at least in intention, have a very different status.

The second point has to do with the class of synthetic *a posteriori* sentences. Is Kant a "reductionist" in regard to this class? In some sense, the answer is "yes." Kant speaks, in fact, of the "derivation" both of empirical concepts and empirical propositions from experience.[44] But to say that Kant is in some sense a reductionist in this regard is of little consequence. For, as we shall see, the very possibility of "deriving" empirical concepts and propositions from experience depends on the possibility of experience in the first place, and *this* possibility depends on certain other propositions being themselves irreducible. Moreover, a point to be made in more detail in the next section of this chapter, not all synthetic *a posteriori* sentences are in a crucial way *equivalent* to "statements about actual and possible sensory states," since at least some of the former have an *objective* character, guaranteed by the Categories, that the latter do not. Kant is quite definitely not a phenomenalist.[45]

The third point I want to make about Quine's thesis is of a more general character. The key part of the thesis, and the one that bears most directly on Kant's position, is that no *viable* distinction can in any case be made between analytic and synthetic sentences, quite independent of the ties between that distinction and the dogma of reductionism. Kant's distinction between analytic and synthetic sentences can be made to turn, I have suggested, on the concept of a possible world. Quine's underlying argument can be reconstrued in the same way: on this reconstruction, his claim is that we cannot say what a possible world must be like. At different times and for different purposes, the limits of the possible may ebb and flow since, according to Quine, we are free to revise even our fundamental logical laws or change the meaning of our basic concepts if the situation warrants. In parallel fashion, we cannot say what a "really possible" world must be like either (granting for the moment that Quine would condone a dis-

[44] See, for example, the first *Critique*, A85/B117.

[45] Again, see Bird, *Kant's Theory of Knowledge*, chapter 1.

tinction between possibility and real possibility), for again the framework of very general scientific-metaphysical principles with which we confront the world and in terms of which a particular description of the class of "really possible" worlds might be framed is subject to change. Kant, on the other hand, clearly thinks that the laws of logic are not subject to change,[46] and directs his main argument to the conclusion that one conception of what it is possible for us to experience—that is, a conception of what constitutes a "really possible" world, a conception embodied in the Categories—is in some deep sense necessary.

There are two different issues at stake. Quine argues that any analytic/synthetic distinction is relative to a set of framework principles, logical and natural laws, etc., themselves subject to change. On the one hand, sentences construed as analytic or synthetic with respect to one set of framework principles will not necessarily be so construed with respect to another. On the other hand, there are no principles that are immune to revision in the face of recalcitrant experience; any sentence of a theory, including those which exhibit its logical and mathematical apparatus, may be revised in the attempt to realign them with the evidence. Schematized, the argument is that the analytic/synthetic distinction implies the non-revisability of certain principles, but every principle is (in principle) revisable, therefore the analytic/synthetic distinction must be given up.

Kant appears to have accepted the implication. That there are certain necessary propositions that can be known *a priori* is coupled by him with the claim that these propositions are not revisable. But one could surely separate the two claims. Relative to a particular set of framework principles we can distinguish analytic, synthetic, and synthetic *a priori* sentences, whether or not these principles might eventually be revised. In fact, Kant's main claim is that certain judgments are syn-

[46] Perhaps simply because the notion of consistency, and hence of what it is to be a possible world, which the laws of logic define, is identical with what is "thinkable," and thus limited in a brute-fact way by conceptual abilities on our part themselves not subject to change.

thetic *a priori* with respect to a given body of knowledge. He has no separate argument—over and above the suggestions of his quasi-psychological view that our minds have an inherent structure—for claiming that these same judgments are not revisable. Despite the attention that has subsequently been devoted to it, revisability *per se* is not the issue on which Kant's case turns. Moreover, even if the judgments he lists as synthetic *a priori* have since been revised or rejected, all that follows is that he was wrong about *these* particular judgments, and not that there are not judgments that play the role he assigns to them.[47] Thus, Quine's argument only partially applies to Kant in that, even if we grant Quine's point, it does not show that certain principles cannot play the kind of role Kant assigns them.

There is, however, a deeper difficulty than the discussion of Quine has so far disclosed. At the heart of his position, and implicit in my elaboration of the concept of a really possible world, is Kant's thesis that the conditions of the possibility of experience are at the same time the conditions of the possibility of the *objects* of experience; a characterization of a really possible object is, therefore, at the same time a characterization of a really possible world, i.e., a world that we are capable of experiencing.

Now in many respects, the development of scientific thought since the 18th century has undermined Kant's views on a variety of questions. Problems that this position, as expressed, for example, in the *Metaphysical Foundations of Natural Science* (1786), is designed to solve are no longer, or no longer the same, problems, and the range of possible solutions to these problems—e.g., continuity, action at a distance, absolute space, the existence of a void, the reality of inertial and gravitational forces—has shifted radically. It is to these kinds of shifts that Quine draws our attention and it is the fact that they have occurred that makes so many of Kant's views vis-à-vis physics seem so dated.

At the same time, I contended a few paragraphs back that

[47] One of the referees has helped me to see this point more clearly.

the development of scientific thought since the 18th century did not undermine Kant's main claim concerning the possibility of synthetic *a priori* knowledge. This must be qualified. One recent strand in the development seems to strike at the heart of Kant's thesis itself: it is at the very least problematic whether in connection with quantum physics, wherein statistical laws and indeterminacy play a major role, the conditions of the possibility of experience—i.e., our experience of the macroscopic world, adequate for the unity of consciousness—are also the conditions of the possibility of the objects of experience, positrons for example. Jules Vuillemin, in his excellent book *Physique et métaphysique kantiennes*,[48] sets the problem this way: "Comment préserver l'unité de la connaissance sans faire violence à la dissociation de l'experience vulgaire et de l'experience scientifique." Kant's achievement was to argue that the conditions of scientific experience, by and large those employed in a Newtonian conception of the world, were at the same time the conditions of common everyday experience, and that they were required as well by the unity of consciousness. Whether an argument of this type is still possible, in the light of the scientific developments since Kant, remains an open question.[49]

presuppositions and truth values

My second reason for calling Kant an "anti-reductionist" has more to do with the rationalist than with the empiricist aspects of his position. There are passages like this one in the first *Critique* at A127:

"However exaggerated and absurd it may sound, to say that the understanding is itself the source of the laws of nature, and so of its formal unity, such an assertion is none the less correct, and is in keeping with the object to which it refers, namely, experience . . . all empirical laws are only special determinations of the pure laws of the understanding,

[48] Paris: Presses Universitaires, 1955, p. 360.

[49] See Wilfrid Sellars, "Philosophy and the Scientific Image of Man," in *Science, Perception and Reality* (New York: Humanities Press, 1963).

under which, and according to the norm of which, they first become possible."

There is also the widely received belief that Kant's principal object was to guarantee Newtonian physics and Euclidean geometry as necessarily descriptive of our experience.[50] In the third and fifth chapters of this book, I will have something to say in detail about Kant's relationship to Euclid and Newton respectively. Here I want only to begin undermining the rationalist interpretation of Kant's position.

To begin with, there are a variety of passages in which Kant explicitly rejects the idea that all the propositions of physics can be derived from *a priori* first principles, even synthetic *a priori* ones. For example, in the first edition of the *Critique*, again at A127: "Certainly, empirical laws as such can never derive their origin from pure understanding. That is as little possible as to understand completely the inexhaustible multiplicity of appearances merely by reference to the pure form of sensible intuition." In the second, at B165: "Pure understanding is not, however, in a position, through mere categories, to prescribe to appearances any *a priori* laws other than those which are involved in a *nature in general*, that is, in the conformity to law of all appearances in space and time. Special laws, as concerning those appearances which are empirically determined, cannot in their specific character be *derived* from the categories, although they are one and subject to them. To obtain any knowledge whatsoever of these special laws, we must resort to experience."

Some commentators have found it difficult to reconcile this sort of passage with those in which Kant insists on the "law-giving" character of the understanding, and hence have con-

[50] A belief apparently reinforced by the proceedings of the *Metaphysical Foundations of Natural Science* (occasionally abbreviated in the footnotes to *MFNS*) and the notes in the *Opus Posthumum* entitled the "Transition from the Foundations of Natural Science to Physics." In this latter work, for example, Kant seems intent on deducing, from *a priori* principles, the system of possible forces in nature, the range of basic properties of matter, and finally, the existence of the ether. From the metaphysics of experience, it appears, one can infer the physics of the natural world.

cluded that the latter represent an earlier stage in his thought that eventually gives way to the mature position in which the sensibility, and the having of empirical intuitions, play as important a role as the understanding in the acquiring of scientific knowledge. This conclusion finds support in the fact that most of the "law-giving" passages were altered or removed by Kant in the second edition of the *Critique*. But it is not necessary to retreat to a biographical thesis to save him from inconsistency. The "law-giving" passages, correctly understood, reconcile with his claim that we cannot, as he puts it at A171/B213, anticipate general natural science.

But there is another aspect of Kant's position that gives even greater support to the rationalist interpretation of it. This aspect has to do with Kant's account of his own procedures, in particular with what he takes to be the premises and conclusion of his central argument. There is, in fact, very general disagreement about what Kant's procedure, especially in the *Critique of Pure Reason*, is. But it has seemed to some commentators that, beginning with the validity of Newtonian mechanics and Euclidean geometry, or perhaps simply the truth of *some* established body of mathematics and physics, he proceeds to search for more general and ultimate principles from which they may be derived. In this same sense, these principles constitute the conceptual framework that makes experience possible.[51]

This interpretation is grounded in Kant's account of his method in #4 of the *Prolegomena*. He there contrasts analytic and synthetic ways of proceeding. On the former, one takes a body of knowledge as given and then "ascends to the sources which are not yet known, and which, when discovered, will not only explain what we know already, but will also exhibit

[51] See, for example, Gottfried Martin, *Kant's Metaphysics and Theory of Science* (Manchester: Manchester University Press, 1955), pp. 79-80: "A set of axioms only has meaning in relation to a system of propositions, and a 'proof' of a set of axioms is only possible in the sense of showing that it is an axiom-set for the system of propositions in question. In the same way the Principles (of pure understanding) are conceived by Kant as principles of possible experience, or quite concretely, the Principles are conceived as the axioms of classical physics."

a large extent of knowledge which springs exclusively from these same sources." On the latter way of proceeding, one takes certain propositions about "pure reason itself . . . both the elements and the laws of its pure employment" and then proceeds to show that, given these propositions, some specified body of knowledge is possible. Thus, the analytic and synthetic methods are also "regressive" and "progressive" respectively; they proceed in different "directions." As one commentator, Robert Paul Wolff, summarizes it: "a regressive method seeks out the presuppositions of a given proposition or body of knowledge," a progressive analysis "begins with the presuppositions and advances to the proposition or body of knowledge."[52]

Kant takes over the notion of analytic and synthetic *methods* from a long-standing tradition in geometry.[53] In that tradition, one proceeds analytically when one begins with a theorem and "works backwards," so to speak, to the axioms of geometry from which it can be derived. To proceed synthetically, on the other hand, is to begin with the axioms and proceed to the demonstration of individual theorems. On analogy with this geometrical tradition, we might be led to picture Kant as beginning the *Prolegomena* with pure mathematics and natural science—that is, bodies of incontestably synthetic *a priori* propositions—and then "ascending" to the principles from which pure mathematics and natural science can be deduced. If we take the same analogy seriously, then in the *Critique of Pure Reason* Kant begins by "inquiring within pure reason itself" and proceeds from the principles that inquiry has disclosed to the demonstration of mathematics and science. The demonstration shows, again according to R. P. Wolff, "*how geometry is possible*, by exhibiting in sys-

[52] *Kant's Theory of Mental Activity* (Cambridge: Harvard University Press, 1963), p. 45.

[53] See, for example, the extremely interesting summary by Pappus (c. 300) in Ivor Thomas, ed., *Greek Mathematical Works* (Cambridge: Harvard University Press, 1941), II, pp. 597-601. For more on the history of the analytic method in particular, see Jaakko Hintikka, "An Analysis of Analyticity," in P. Weingartner, ed., *Description, Analytizität, und Existenz* (Salzburg and Munich, 1966).

tematic form a set of conditions *from which* geometry can be deduced."[54]

I emphasize all this because many commentators seem to construe Kant's use of analytic and synthetic methods in just this way.[55] On their view, Kant deduces the content of physics and (much less plausibly) mathematics from general synthetic *a priori* principles.[56] These latter are thus "presupposed" in the sense that they are sufficient conditions, or part of a set of sufficient conditions, for mathematics and physics. When Kant says that the Transcendental Deduction is a search for the "*a priori* conditions of the possibility of experience" he means, on this view, that it is a search for the premises of a deductive argument. In this sense, we say that classical mechanics presupposes the calculus, and any science, or body of deductively linked propositions, presupposes logic. Again, this sense seems to accord well with Kant's remark that the *Critique*, on analogy with logic, is a "propadeutic" to the sciences.

Nevertheless, I think that this is a mistaken interpretation.

[54] *Kant's Theory of Mental Activity*, p. 45. As one of the referees has insisted, this assertion cannot really be taken seriously as it stands. For in the *Critique*, Kant nowhere attempts to prove any particular mathematical propositions. What he tries to do is to guarantee the "objective validity" of mathematics, but this does not involve a demonstration of the type indicated. The situation is not so clear with respect to the basic propositions of physics. See chapters 3 and 5 for detailed elaboration of these claims.

[55] Wolff goes so far as to say that "there are really two and only two plausible interpretations of the *Critique*. Either Kant begins by assuming the validity of mathematics and natural science, and then works backward to the *a priori* sufficient conditions of their validity. Or else he begins with some premise which is universally granted, and works forward by a synthetic deductive proof to the validity of mathematics and natural science." *Ibid.*, p. 54.

[56] T. D. Weldon claims that "pure formal logic" provides the requisite first principles for Kant (although Weldon thinks the deduction fails). See his *Kant's Critique of Pure Reason*, 2nd edition (Oxford: Clarendon Press, 1958), pp. 170ff. But this could hardly be the case; pure formal logic defines merely analytic, not synthetic *a priori* knowledge, and if our earlier claim was correct, no synthetic propositions can be derived from analytic ones. This is not to say that the principles of pure formal logic did not suggest, by analogy, principles of "transcendental logic" that are synthetic *a priori*.

My direct reasons may be set out briefly. In the first place, if the relationship between, for example, physics (even of its "pure" part) and the Categories is supposed to be deductive, then Kant's argument in the *Critique of Pure Reason* is almost totally unsuccessful. There just is no "deduction" in this sense.[57] In the second place, this interpretation conflicts with other passages in the *Critique*, as we shall see in a moment. In the third place, the fact that certain principles serve as premises in a deductive argument does nothing to establish Kant's intended conclusion that it is *only* on the assumption of such principles that the possibility of experience may be shown, that is, that they are in some sense "necessary." A great many different sets of sufficient conditions serve for the derivation of mathematics and physics. My indirect reason for resisting this interpretation, however, is the one on which I place more weight. It is that an alternative interpretation, to be set out shortly, much more adequately represents Kant's position.

A second and, I believe, equally unacceptable interpretation of Kant's procedure is that the "*a priori* conditions of the possibility of experience" are not sufficient but rather *necessary* conditions.[58] In this sense, p presupposes q not where q implies p, but rather just in case p implies q. Kant often refers, in fact, to the "*a priori* conditions of the possibility of experience" as "necessary conditions," and there are passages where he suggests that natural science and mathematics figure not as the conclusion of the argument but as the premises. For example, in #4 of the *Prolegomena*:

"It is fortunately the case that, although we cannot assume that metaphysics as a science is *real*, we can confidently say that certain pure synthetic knowledge *a priori* is real and given, namely *pure mathematics* and *pure natural science*. . . . We have therefore some, at least incontested, synthetic knowl-

[57] It might be noted in passing that Kant's use of the word "deduction" in expressions like "Transcendental Deduction" is in what he indicates is the *legal*, in contrast with the *logical*, sense. See the *Critique*, A84/B116. In this sense, "deduction" is roughly synonymous with "justification."

[58] Arthur Pap offers this interpretation in his book, *The A Priori in Physical Theory* (New York: Columbia University Press, 1946).

edge *a priori*, and do not have to ask whether such knowledge is possible (for it is real), but only *how it is possible*, in order to be able to deduce from the principle of the possibility of the given knowledge the possibility of all other synthetic knowledge *a priori*."

A major difficulty with this interpretation is that, once again, Kant nowhere attempts a "deduction" of the Categories from physics or mathematics, even from their "pure" parts, nor is it easy to see how one might be carried out. This fact, taken together with passages in which Kant suggests that an analytic or regressive argument can lead us to "necessary conditions," has led some commentators to cast around for another sense for "necessary conditions." At least two offer themselves as candidates. The first is that a necessary condition is a sufficient condition, "necessary" in the sense that it is the *only* sufficient condition.[59] There are places where Kant does appear to be proceeding in approximately this manner. The Transcendental Aesthetic, for instance, takes as a premise that only certain views of space are possible: either space is relational or it is not. For certain reasons, only one of these views is tenable. Therefore, this view of space is a necessary as well as a sufficient condition for the synthetic *a priori* status of geometry. More generally, Kant sometimes begins with a pair of alternatives delimited by the law of excluded middle, and then tries to show that only one of the alternatives accounts for the knowledge we do in fact have.

The other meaning that can be given to "necessary condition" is that the proposition taken to be a necessary condition is itself a necessary proposition. And again, Kant occasionally seems to proceed by showing that the conditions arrived at by way of a regressive argument are themselves necessary, sometimes because their negation is inconceivable, sometimes, particularly in certain first-edition sections of the Transcendental Analytic, because they describe "brute-fact" aspects of the understanding.

[59] See the *Prolegomena*, Academy edition, IV, p. 276n., where the analytic method is described as "starting from what is sought, as if it were given, and ascending to the *only* conditions under which it is possible" (my italics).

Both of these alternative views of "necessary condition" throw light on the text, but they are subject to the same basic difficulties as the primary sense, difficulties indeed to which any deductive interpretation of "presupposition" is subject. The propositions of pure mathematics and pure natural science do not imply nor are they implied by the Categories. Moreover, there is, as already mentioned, another interpretation of "presupposition" that makes a good deal more systematic sense of Kant's position.[60] As was the case with the concept of a possible world, it is an idea introduced from the outside. But, as before, its introduction can be justified eventually by an appeal to the text.

This interpretation of "presupposition" has been developed by Bas van Fraassen,[61] although not specifically in connection with Kant. I will sketch it first, and then try to justify taking it as Kant's position rationally reconstructed.

For van Fraassen, "presupposition" is a *semantic* relation and is characterized as follows:

(1) *A presupposes B* if and only if *A* is neither true nor false unless *B* is true; or, what comes to the same thing:

(2) *A presupposes B* if and only if (a) if *A* is true, then *B* is true, (b) if *A* is false, then *B* is true.

In the paradigm case, the presupposition at stake is existential. We might put the point in terms of the dictum that a property cannot be truly or falsely attributed to what does not exist. Thus, to use the well-worn example, "The King of France (in 1977) is bald" is neither true nor false, simply because the King of France does not exist.[62]

This characterization of "presupposition" has two key fea-

[60] Still other interpretations of the presupposition relation are suggested by Arthur Pap in "Does Science Have Metaphysical Presuppositions?" in H. Feigl and M. Brodbeck, eds. *Readings in the Philosophy of Science* (New York: Appleton-Century-Crofts, 1953).

[61] On a suggestion of Karel Lambert's, who suggested to me that it might be used in a reconstruction of Kant's position. See van Fraassen's paper, "Presupposition, Implication, and Self-Reference," *Journal of Philosophy*, LXV (1968), pp. 136-152.

[62] See P. F. Strawson's paper, "On Referring," *Mind* (1950), pp. 320-344, where this way with the case is first taken.

tures. The first is that, in van Fraassen's words, "presupposition is a trivial semantic relation if we hold the principle of bivalence (that every sentence is, in any possible situation, either true or false). In that case, every sentence presupposes all and only the universally valid sentences." This is perhaps an obvious consequence of the characterization.

The second feature is that presupposition must be distinguished from implication. To show this, it is first necessary to define negation: the negation of a sentence A is true (false) just in case A is false (true). Now we can characterize presupposition further as follows:

(3) *A presupposes B* if and only if (a) if A is true, then B is true, (b) if (not-A) is true, then B is true.

It should be clear that on this characterization, the analogue to *modus tollens* for presupposition, which is valid for implication, does not hold, *viz.*

$$\frac{\begin{array}{r} A \text{ presupposes } B \\ (\text{not-}B) \end{array}}{\text{therefore, } (\text{not-}A)}$$

Further, (3a) shows that

$$\frac{\begin{array}{r} A \text{ presupposes } B \\ (\text{not-A}) \end{array}}{\text{therefore, } B}$$

is valid, whereas the analogue does not hold for implication. On the other hand, if A either presupposes or implies B, the argument from A to B is valid.[63]

The first question we have to ask is whether Kant does give up bivalence. If he does not, this notion of presupposition will be useless. As a preliminary, bivalence—every proposition is

[63] R. G. Collingwood has made a great deal of the notion of presupposition, both in commenting on Kant and in developing his own philosophical position. It is difficult to say whether or not he uses "presupposition" in the way just sketched, however, largely because he illustrates it with a bewildering variety of examples. See *An Essay on Metaphysics* (Oxford: Clarendon Press, 1940), especially pages 290-291.

either true or false—must be distinguished from the law of contradiction—no proposition can be true and false at the same time. There seems, indeed, to have been a certain amount of blurring between these two principles in the tradition, and little appreciation of the fact that they are independent or of the possibility of maintaining that at least some propositions are neither true nor false. Leibniz, for example, perhaps more sensitive to these issues than anyone else, says in the *Nouveaux Essais*, Livre IV, chapitre II, #1, that "the principle of contradiction" contains two principles: that a proposition cannot be true and false at the same time (what we are calling the principle of contradiction) and that every principle is either true or false (the principle of bivalence).[64] But he is indifferent to the differences between them, even failing to provide each with its own name.[65]

Retention of the principle of bivalence has the following corollary for Leibniz. All (atomic) sentences[66] in which the subject term fails to refer are false, a view made prominent in our own time by Russell. Perhaps most explicit is fragment #393 in Couturat's collection, *Opuscules et Fragments inédits de Leibniz*: "In order, namely, to keep (the principle) that every proposition is true or false, (I consider) as false every proposition that lacks an existent subject or real term."[67] I do not know if Kant has a general account of propositions containing non-referring subject terms. The propositions in which he is most interested, in any case, are not those containing such terms, e.g., "Pegasus," but rather those containing subject terms, like "God" and "soul," for which no corresponding

[64] *Nouveaux Essais sur l'Entendement Humain* (Paris: Garnier-Flammarion, 1966).

[65] Bertrand Russell, commenting on Leibniz on the law of contradiction, seems similarly unaware of the distinction: "the law states simply that any proposition must be true or false, but cannot be both." *The Philosophy of Leibniz*, second edition (London: Allen & Unwin, 1937), p. 22.

[66] The restriction to sentences without logical structure seems required by the law of excluded middle and the existence of true negative existentials.

[67] For a discussion of this and related items in Leibniz, see Benson Mates, "Leibniz on Possible Worlds," pp. 514ff., whose reference to Couturat I quote here.

perceptible object or sensible intuition can be given.[68] It is Kant's view, I contend, that *these* propositions are neither true nor false *for us*.

In a way, then, if I am correct, the first presupposition for Kant is an existential presupposition. But the existential presupposition is given an epistemological twist. It is not that for a proposition formulated, say, in first-order logic, to have a truth value that the singular terms and bound variables it contains must have a reference, but that the objects to which the singular terms and bound variables purport to refer are capable of being experienced by us. They must be knowable objects. But the only sorts of objects we can experience, Kant goes on to argue, are those, in particular, capable of occupying determinate spatial-temporal position. In other words, the only kind of intuition of objects of which we are capable is sensible intuition. Thus, that certain sorts of statements have truth values for us, or as he says have "objective reality," presupposes space and time as *a priori* conditions of experience.

Kant sometimes makes the same point in a negative way. If we were capable of "intellectual intuition," intuition of objects not limited by spatial-temporal conditions, then the presuppositions of, for example, rational theology, notably "God exists," would have a truth value. In the *Inaugural Dissertation*, when Kant (still following Leibniz) assures us that we are capable of such intellectual intuition, propositions of this kind are given truth values. But in the *Critique of Pure Reason*, the doctrine of intellectual intuition is given up, with the result that "all categories through which we attempt to form the concept of such an object allow only of an empirical employment, and have no meaning whatsoever when not applied to objects of possible experience, that is, to the world of sense. Outside this field they are merely titles of concepts,

[68] See B302n.: "In a word, if all sensible intuition, the only kind of intuition we possess, is removed, not one of these concepts can in any fashion *verify* itself, so as to show its *real* possibility. Only *logical* possibility then remains, that is, that the concept or thought is possible. That, however, is not what we are discussing, but whether the concept relates to an object and so signifies something."

which we may admit, but through which (in and by themselves) we can understand nothing (A696)."[69]

The words "for us" that I have added to the claim that for Kant certain propositions (by and large those of traditional metaphysics) are truthvalueless crucial. Kant argues that, God's existence is a requirement of practical reason; the coherence of our moral life depends on postulating and believing this proposition. It follows that the proposition is *thinkable* (i.e., self-consistent) and, presumably, has a truth value as well. Kant's point is that the proposition is not *knowable* (i.e., we cannot know whether it, or any other proposition about the supersensible, is true or false). Thus we are to understand the truth valueless thesis in an epistemological way, with reference to our *knowledge* of the truth values in question. "*p* is true" is to be understood as "we can know that *p*" and "*p* is false" as "we can know that not-*p*."[70]

There are a number of places in the *Critique* where the principle of bivalence (epistemologically interpreted) is in question. None is more important than the discussion of the first and second Antimonies of Pure Reason. On the presupposition that the world as a totality or completed series is a possible object of experience, we are led to consider the following disjunctions: either the world has a beginning in time or it does not; either the world is limited in space or it is not. Kant wants to say that neither member of these pairs of apparently contradictory propositions has a truth value, since the presupposition in question is not satisfied; "worlds" are not really possible objects. What he does in fact say is that both

[69] See also A222/B270 in the section entitled the "Postulates of Empirical Thought:" "A substance which would be permanently present in space, but without filling it . . . , or a special mental power of intuitively anticipating the future . . . , or lastly a power of standing in a community of thought with other men, however distant they may be—are concepts the possibility of which is altogether groundless, as they cannot be based on experience and its known laws; and without such confirmation they are arbitrary combinations of thoughts, which, although indeed free from contradiction, can make no claim to objective reality, and none, therefore, as to the possibility of an object such as we have professed to think."

[70] This paragraph incorporates suggestions made by two of the referees.

members of these pairs of propositions are false, although he adds at once that in such cases we cannot conclude from the falsity, even the absurdity, of the one to the truth of the other. But calling them "false" is hardly consistent with other aspects of Kant's position. For one thing, there are parallel cases of failure of epistemological presupposition, the case of "God exists," for example, where Kant quite definitely does not want to say that as a result the proposition in question is *false*. For another thing, it is Kant's position that when there is a failure of presupposition we cannot *know* whether the proposition in question is true or false, which for him is tantamount to saying that at least *for us* it has no truth value. What is consistent with Kant's position, and what I argue is implicit in it, is the abandonment of bivalence. Abandoning bivalence as well achieves Kant's aim in his discussion of at least the first two Antimonies, to avoid having to admit that, on logical grounds alone, one or the other member of these pairs of suspect propositions must be true.

Interestingly, there is a statement of a principle[71] that very much resembles the characterization of presupposition I have sketched and that comes at the end of the discussion of the Antimonies, at A503/B531: "If two opposed judgments presuppose an inadmissible condition, then in spite of their opposition, which does not amount to a contradiction strictly so-called, both fall to the ground, inasmuch as the condition, under which alone either of them can be maintained, itself falls." The "inadmissible condition" is, once again, that the object in question is not one we are capable of experiencing. If a judgment and its negation presuppose such a condition, they are not simply "false"; rather, they "fall to the ground" (*sie fallen weg*). The presupposition in such cases fails.

[71] Kant takes it to be a statement of the medieval maxim: *non entis nulla sunt predicata* (nothing can be predicated of what does not exist). There are other, slightly varying, statements of the same principle in the *Critique of Pure Reason*, A793/B821 (in the course of Kant's most extended discussion of the nature of *proofs*), in the long letter to Marcus Herz of February 21, 1772, and in #52b of the *Prolegomena*. I discuss this principle as it is interpreted by Kant and Leibniz, in part recapitulating paragraphs above, in my paper, "Non entis nulla sunt attributa" in the *Proceedings* of the fourth international Kant Congress (*Kant-Studien*, 1974).

This is obviously not the end of the matter. The interpretation just broached needs a good deal more elaboration and support. I will try to provide both in detail. But there are two more points that might be made here that connect the earlier discussion of the synthetic *a priori* with the present discussion of "presupposition." The first is simply this. Synthetic *a priori* judgments determine what is "really possible"; they do not allow us to distribute definite truth values to individual purely synthetic propositions. They allow for the "possibility" of these propositions in the sense that, specifying the conditions that have to be satisfied for their subject terms to refer to really possible objects, they guarantee their having a truth value, i.e., their meaningfulness, i.e., their objective reality. It is in this way that they are "conditions of the possibility of experience" or ultimate presuppositions. If, as Quine suggests, we were to say that certain propositions, formerly held recalcitrant to experience, must be revised, then certain other propositions, formerly held meaningless, would acquire a truth value. Similarly, if we were to change our presuppositions, change the limits of the "really possible," rational theology, to choose but one example, might become "possible" in the sense I am attributing to Kant. Kant's argument will be that for some very deep-rooted reasons these presuppositions cannot be changed.

The other, closely related point, is that it is only insofar as they are presupposed by experience that the Categories are themselves really possible. If the Categories were "borrowed directly from experience," then the issue of their real possibility would not arise. For in that case the perceptibility of their corresponding objects would be guaranteed. But the Categories are not borrowed directly from experience. They are *a priori*. Similarly, if the Categories were analytic, then the issue of their real possibility would not arise, at least not in quite this form.[72] For as analytic their truth (in some sense) in

[72] The same sort of point is made earlier on in the discussion of the Antimonies, at A485/B513: "If our question is directed simply to a yes or no, we are well advised to leave aside the supposed grounds of the answer, and first consider what we should gain as the answer is in the affirmative or in the negative. Should we then find that in both cases the outcome is sheer non-

all possible worlds, *a fortiori* in all really possible worlds, would be guaranteed. But the Categories are not analytic.[73] They are synthetic. It is precisely Kant's anti-reductionism, in fact, that leads to *the* Critical question: how are synthetic *a priori* judgments possible?[74] On my reading, this comes to asking how it is that synthetic *a priori* judgments have truth values, i.e., how can the existence of experienceable objects corresponding to their subject terms be guaranteed? Without experienceable objects corresponding to their subject terms, without a proof thereof, the Categories become, like the judgments of traditional metaphysics, meaningless. Kant's answer is to argue, transcendentally, that the *object* in question is the whole of experience, and that the real possibility of the Categories can be established *a priori*, as conditions of possible experience in general.

---

sense, there will be good reason for instituting a critical examination of our question, to determine whether the question does not itself rest on a groundless presupposition. . . ."

[73] See the next chapter for a more extended discussion of the *meaningfulness* of analytic judgments.

[74] Thus, the problem he faced grew directly out of his rejection of his philosophical predecessors.

Chapter 2: Kant's philosophy of mathematics

ONE well-entrenched view of Kant's philosophy of mathematics is as follows. Kant took Euclidean geometry as the paradigm of mathematical reasoning. However, neither in Euclid's own time nor in Kant's could all of Euclid's proofs be carried out without the use of geometrical constructions. So Kant was led to insist on the centrality of constructions or, in his preferred vocabulary, "intuitions" in mathematical reasoning.[1] Yet further developments—principally the formalization techniques associated with the name of Hilbert—reveal that constructions are inessential from a logical point of view. Euclidean geometry can be given a formal axiomatization. Moreover, Kant's emphasis on the synthetic *a priori* character of mathematical propositions stems from his failure to distinguish clearly, as his successors have done, between pure and applied mathematics. Pure mathematics is analytic, hence *a priori* as well, but applied mathematics is synthetic. Therefore, Kant's views are at best out-moded and old-fashioned. They confuse accidental and essential features of mathematical reasoning, questions of form with questions of content. As a not otherwise unsympathetic commentator recently put it: Kant's views about mathematical method "though ingenious, have been revealed by later work on the philosophy and logic of mathematics as thoroughly and tiresomely wrong."[2]

[1] Thus Bertrand Russell, *Introduction to Mathematical Philosophy* (London: Allen & Unwin, 1919), p. 145: "Kant, having observed that the geometers of his day could not prove their theorems by unaided argument, invented a theory of mathematical reasoning according to which the inference is never strictly logical, but always requires the support of what is called 'intuition.' "

[2] Jonathan Bennett, *Kant's Analytic,* p. 4. See also G. T. Kneebone, *Mathematical Logic and the Foundations of Mathematics* (Princeton: D. Van Nostrand Company, 1963), p. 249: "Kant's conception of mathematics has long been obsolete. . . ."

The preceding sketch is not completely misleading. There is some basis for it both in Kant's text and in the history of mathematics. But, at the same time, the sketch obscures much of what is interesting in Kant's view. In particular, it does not enable us to understand many of the things Kant says about arithmetic and algebra, or to see what connection there might be between the use of constructions in carrying out geometrical proofs and the synthetic *a priori* character of sentences like "7 + 5 = 12." Besides, the entrenched view embodies a rather facile refutation. I contend that Kant's philosophy of mathematics is more than a mere historical curiosity.[3]

One reason why Kant's position has not been better understood is that the importance of the remarks he makes near the end of the *Critique of Pure Reason*, in the first chapter of the Transcendental Doctrine of Method (A713/B741ff.) have not been generally appreciated. These remarks, as E. W. Beth[4] and then K.J.J. Hintikka[5] first insisted, throw a great deal of light on what Kant has to say about mathematics in the second preface to the *Critique*, the Introduction and the Transcendental Aesthetic—particularly so if we accept Beth's thesis that these remarks are logically, if not also chronolog-

[3] On grounds rather different from those of the Dutch intuitionists, whose position is so well known I have not commented on it here. See A. Heyting, *Intuitionism: An Introduction* (Amsterdam: North-Holland Publishing Company, 1965), pp. 41ff.

[4] For Beth's reconstruction of Kant's position, see his "Über Lockes 'allgemeines Dreieck,' " *Kant-Studien*, 48 (1956–57), pp. 361–380, and *The Foundations of Mathematics* (Amsterdam: North-Holland Publishing Company, 1965), pp. 41ff.

[5] Among Hintikka's many papers on Kant's philosophy of mathematics, see "Kant's 'new method of thought' and His Theory of Mathematics," *Ajatus*, 27 (1965), pp. 37–43, "Are Logical Truths Analytic?" *Philosophical Review*, 74 (1965), pp. 178–203, "Kant on Mathematical Method," *The Monist*, 51 (1967), pp. 352–375, "An Analysis of Analyticity," "Are Logical Truths Tautologies?" "Kant Vindicated," and "Kant and the Tradition of Analysis," in Paul Weingartner, ed., *Description, Analytizität, und Existenz* (Salzburg and Munich, 1966). The first three of these papers are reprinted in *Knowledge and the Known* (Dordrecht: D. Reidel Publishing Company, 1974), the last four in *Logic, Language-Games and Information* (Oxford: Clarendon Press, 1973).

ically,[6] prior to the theory of mathematics expounded in the Aesthetic.

In the opening sections of the *Critique*, on which the view I sketched is principally based, it is extremely difficult to say what *argument* Kant is setting out for his claim that the propositions of mathematics are synthetic. For example, he tells us that

"We might, indeed, at first suppose that the proposition $7 + 5 = 12$ is a merely analytic proposition, and follows by the principle of contradiction from the concept of a sum of 7 and 5. But if we look more closely we find that the concept of the sum of 7 and 5 contains nothing save the union of the two numbers into one, and in this no thought is being taken as to what that single number may be which combines both. The concept of 12 is by no means already thought in merely thinking this union of 7 and 5; and I may analyze my concept of such a possible sum as long as I please, still I shall never find the 12 in it. . . . That 5 should be added to 7, I have indeed already thought in the concept of a sum $= 7 + 5$, but not that the sum is equivalent to the number 12. Arithmetical propositions are therefore always synthetic" (B15-16).

Confronted by such passages, generations of students have felt called upon to perform a *gedanken-experiment*, trying to determine whether in fact one thinks the concept 12 in thinking the union of 7 and 5. But the results of this introspection will probably always prove inconclusive. Conceptual connections are rarely so readily revealed. Further, Kant's appeal to the use of the fingers in counting has never served to help matters very much. Only later in the *Critique* do the reasons why Kant thinks mathematical propositions are synthetic begin to emerge in a clear, non-psychological form.

the reduction of mathematics to logic: the Frege-Russell program

In calling them "synthetic," Kant first wants to emphasize the

[6] In fact, the Berlin prize essay of 1764, "Enquiry Concerning the Clarity of the Principles of Natural Theology and Ethics," contains much the same sort of view.

irreducibility of mathematical propositions to the fundamental principles of logic: "*All mathematical judgments, without exception, are synthetic.* This fact, though incontestably certain and in its consequences very important, has hitherto escaped the notice of those who are engaged in the analysis of human reason, and is, indeed, directly opposed to all their conjectures. For as it was found that all mathematical inferences proceed in accordance with the principle of contradiction . . . , it was supposed that the fundamental propositions of the science can themselves be known to be true through that principle. This is an erroneous view. For though a synthetic principle can indeed be discerned in accordance with the principle of contradiction, this can only be in another synthetic proposition is presupposed, and if it can then be apprehended as following from this other proposition; it can never be so discerned in and by itself" (B14).

Although he does not mention him by name, Kant evidently has Leibniz in mind. For it was Leibniz's view[7] that all mathematical propositions can be "reduced" to definitions and the principle of contradiction. In short, Leibniz regarded all mathematical propositions as analytic. Kant contends, to the contrary, that although in a mathematical proof the steps follow one another analytically, that is, in accordance with the principle of contradiction, the initial premises of the proof and the conclusion are themselves synthetic.

Two points concerning Leibniz's view should be noted. One is his belief that the principle of contradiction suffices to "demonstrate" all mathematical propositions. It is not clear precisely what this means. It could mean that for Leibniz any argument, hence also one with a mathematical proposition as

[7] Expressed, for example, in the second letter to Clarke. "The great foundation of *mathematics* is the *principle of contradiction or identity*, that is, that a proposition cannot be *true and false* at the same time, and that therefore *A* is *A*, and cannot be *not A*. This single principle is sufficient to demonstrate every part of arithmetic and geometry, that is, all mathematical principles." From the *Leibniz-Clarke Correspondence,* ed. Alexander. See also the *Nouveaux Essais,* Livre IV, chapitre VII, #10, for a sample demonstration, that of "2 + 2 = 4," from definitions of "2," "3," and "4" and the axiom "if equals are substituted for equals, the result is equal."

conclusion, is valid just in case the conjunction of its premises with the denial of the conclusion is self-contradictory. It could also mean that for Leibniz the negation of any individual mathematical proposition is self-contradictory. In the passage quoted at B14, Kant suggests that Leibniz runs these two possible meanings together. The other point is that the reduction of mathematics to logic *chez* Leibniz, using definitions and the laws of logic to prove particular mathematical propositions, does not really amount to more than a series of programmatic remarks. It remained for Frege at the end of the 19th century and Russell at the beginning of the 20th actually to carry out the reduction in detail. In fact, the continuing (although at the present time weakening) rejection of Kant's claims, especially as concerns the synthetic character of arithmetic propositions, is largely based on the Frege-Russell program.[8] The logical apparatus used, moreover, includes other principles than the principle of contradiction. Still, the program continues to have a very Leibnizian formulation. Frege, for example, says on page 4 of *The Foundations of Arithmetic*: "The problem becomes . . . that of finding the

[8] It should be pointed out that Frege, unlike Russell, agreed with Kant about the synthetic character of geometrical propositions. "We shall do well in general not to overestimate the extent to which arithmetic is akin to geometry. . . . One geometrical point, considered by itself, cannot be distinguished in any way from any other; the same applies to lines and planes. Only when several points, or lines or planes are included together in a single intuition do we distinguish them. In geometry, therefore, it is quite intelligible that general propositions should be derived from intuition; the points or lines or planes which we intuit are not really particular at all, which is what enables them to stand as representatives of the whole of their kind. But with numbers it is different; each number has its own peculiarities. To what extent a given particular number can represent all the others, and at what point its own special character comes into play, cannot be laid down generally in advance." *The Foundations of Arithmetic*, translated by J. L. Austin (New York: Harper & Row, 1953), pp. 19–20. I have quoted the passage at length because I will return to Frege's views on geometry in the next chapter, because it throws light on Beth's reconstruction of Kant's position, to which I will turn shortly, and because I think it explains Kant's remark in the *Critique* that "the evident propositions of numerical relation are indeed synthetic but are not general like those of geometry, and cannot, therefore, be called axioms but only numerical formulas" (A164/B205).

proof of the proposition, and of following it right back to the primitive truths. If, in carrying out this process, we come only on general logical laws and on definitions, then the truth is an analytic one." And I think that Kant's reasons for claiming that mathematics does not reduce to logic (hence is not analytic) apply as much to the Frege-Russell program as to Leibniz's. A brief summary of the Frege-Russell program should provide a context for the elaboration of these reasons.

The first step in the program is to reduce the various branches of mathematics—e.g., analysis, algebra, and geometry to arithmetic—in part by way of the construction of the real in terms of the rational numbers and of the rational in terms of the natural numbers.

The second step is to show that arithmetic can be reduced to logic. To put it in *very* simple terms, this is accomplished as follows in Russell and Whitehead's *Principia Mathematica.* We presume that arithmetic can be developed on the basis of five axioms set down late in the last century by the Italian mathematician Peano;[9] from them all the properties of the integers can be derived by strictly logical reasoning. The five axioms are:

A.1: *0* is a number.
A.2: The successor any number is a number.
A.3: No two numbers have the same successor.
A.4: *0* is not the successor of any number.
A.5: If P is a predicate true of 0, and whenever P is true of a number n, it is also true of the successor of n, then P is true of every number.

The trick is to define the arithmetical notions that appear in these axioms in logical terms. Once this is done, the reduction of mathematics to logic would seem to proceed without difficulty. Once numbers are construed in logical terms, the basic arithmetical operations—addition and multiplication—can be

[9] In fact, these axioms suffice for only a fragment of arithmetic, but nothing turns on the point.

similarly defined. In this way, (true) mathematical propositions turn out to be provable as theorems in logic.

The three notions involved are "*0*," "is a number," and "is the successor of." In fact, the second can be defined in terms of the first and the third. To say that *n* is a natural number is to say that *n* is *0*, or is the successor of *0*, or is the successor of the successor of *0*, etc. We have only to deal with "*0*" and "is the successor of." "*0*" can be defined as "the set which contains only the set that has no members," that is, the set that contains the null set as its only member. And the successor of any number *n* is, on Frege's famous definition, the set of all sets that, when deprived of a member, come to belong to *n*.[10] Since these defining expressions can in turn be analyzed in terms of "all," "or," "not," and "is-a-member-of," which can be construed, at least provisionally, as *logical* expressions, the reduction of mathematics to logic is virtually complete.[11]

the Beth-Hintikka reconstruction

Kant begins his remarks at A713/B741 of the *Critique of Pure Reason* with a general contrast between mathematical and philosophical knowledge. He says there that "*Philosophical knowledge is the knowledge gained by reason from concepts*; mathematical knowledge is the knowledge gained by reason from the *construction* of concepts." Further, it is in virtue of its "constructive" character that mathematics is synthetic.

What is it to construct a concept? Kant answers that "To construct a concept means to exhibit *a priori* the intuition which corresponds to the concept." For the moment, let us

[10] In fact, this is Frege's definition as simplified by Russell. See W. V. Quine, *Set Theory and Its Logic* (Cambridge: Harvard University Press, 1963), #12, for a discussion of Frege's definitions and a comparison of them with the alternative ways of construing numbers proposed by von Neumann and Zermelo.

[11] Compare Leibniz's remark in his *Universal Mathematics*: "Arithmetic and Algebra can be so treated by means of Logic, as if they were Logical Mathematics, so that in this way Universal Mathematics coincides in effect with Logistics and the Logic of Mathematics; hence our Logistics is given in some places the name, Mathematical Analysis."

forget about what "a priori" might mean here and concentrate on the notion of exhibiting an intuition corresponding to a concept. For Kant, an intuition *represents* an individual object (although Kant sometimes construes it as *being* an individual object).[12] There is no need to mention all of the relevant passages here.[13] In the opening paragraphs of his lectures on logic,[14] Kant explicitly defines an intuition (*Anschauung*) as an individual representation (*einzelne Vorstellung, representatio singularis*), and in the *Critique of Pure Reason* and elsewhere intuitions are invariably contrasted with general concepts, for example at A320/B376: ". . . representation . . . is either *intuition* or *concept*. The former relates immediately to the object and is single, the latter refers to it mediately by a feature which several things may have in common."[15] Hintikka puts this in a more contemporary way by suggesting that we are to take intuitions in Kant's technical sense as *singular terms*—

[12] Perhaps it is an ambiguity shared with "being a representation."

[13] See Hintikka's paper, "On Kant's Notion of Intuition," in T. Penelhum and J. H. MacIntosh, ed., *The First Critique* (Belmont, California: Wordsworth Publishing Company, 1969), pp. 38–53.

[14] Academy edition, IX, p. 91.

[15] On the basis of passages such as this one, Hintikka seems to me to be correct in insisting that Kant intends "immediately" in a logical rather than an epistemological sense, and hence that the immediacy criterion for intuitions is little more than a corollary of the singularity criterion. This interpretation has been criticized by Charles Parsons, "Kant's Philosophy of Arithmetic," in S. Morgonbesser, P. Suppes, and M. White, ed., *Philosophy, Science, and Method* (New York: St. Martin's Press, 1969), pp. 568–594. See also Hintikka's reply to Parsons, "Kantian Intuitions," *Inquiry*, 15 (1972), pp. 341–345, and the papers by Howell, "Intuition, Synthesis, and Individuation in the *Critique of Pure Reason*," and Manley Thompson, "Singular Terms and Intuitions in Kant's Epistemology," *Review of Metaphysics*, 26 (1972), pp. 314–343, which touch on various aspects of the controversy. If we were to accept Parsons' interpretation—that it is part of the *meaning* of "intuition" that intuitions are quasi-perceptual (and thus that the "immediacy criterion" is independent of the singularity criterion and has epistemological import)—then how would we be able to understand Kant's claim (at B146) that "as the Aesthetic has shown, the only intuition possible to us is sensible"? On Parsons' interpretation, that (human) intuition is sensible follows as a trivial consequence of the definition and should not require the extended argument of the Aesthetic.

symbols that refer to individual objects or are used as if they so referred. The suggestion seems particularly well-suited to Kant's characterization in #8 of the *Prolegomena*: "Intuition is a representation such as would depend directly on the presence of the object."

In constructing a geometrical figure, a triangle say, we often represent it by a figure drawn on a blackboard. In the same way, we "construct" arithmetical or algebraic concepts when we represent individual quantities, perhaps by the fingers of a hand, perhaps by numerals or letters. What is important is not so much the "representative"—that is, the "pictorial" character of such "constructions"—but the fact that in using them we *refer to* or *talk about* individual mathematical objects, whatever these might be.[16]

Hintikka's reconstruction of Kant's philosophy of mathematics proceeds as follows.[17] The underlying theme is that mathematics is "constructive" insofar as its proofs turn on the introduction of intuitions or individual representations. Proofs that introduce such individual representations in the course of being carried out are synthetic, those which do not are analytic. Thus, for example, "a geometrical argument in the course of which no new geometrical entities are 'constructed'—that is, introduced into the discussion—will normally be converted into a quantificational argument in the course of which no new free individual symbols are introduced."[18] Since not all geometrical arguments have such a form, we must say that geometry is, at least in part, synthetic. More generally, many mathematical proofs convert into quantificational arguments in the course of which new free individual symbols[19] are introduced ("concepts are constructed"); for that reason, they are to be considered syn-

[16] Thus we might say, with pun intended, that the constructions are "symbolic."

[17] What follows is an informal and intuitive sketch of what is in Hintikka's papers a formal and elaborately developed view.

[18] "Are Logical Truths Tautologies?" Weingartner, p. 201.

[19] To use Hintikka's own terminology. It is perhaps more standard to speak of argument constants.

thetic. By extension, we call a sentence analytic if it is the conclusion of an argument the respective steps of which do not introduce new individuals into the discussion; otherwise it is synthetic.

This theme can be made somewhat more precise by way of variations on it with respect to quantification theory. One such variation is that an analytic argument cannot lead from an assertion about one individual to an assertion about another individual. For Kant, according to Hintikka, inter-individual inferences concerning existence are impossible by analytic means.[20] But we can show that many of the truths of quantification theory turn on inter-individual inferences concerning existence.[21] Thus, these truths are to be taken, in a sense that is primary for Kant, as synthetic. A second variation on the underlying theme is that an argument step is analytic if it does not increase the number of individuals we are considering together in their relation to each other, where the number of individuals considered together in a sentence *S* is the number of free singular terms of *S* plus the maximal number of quantifiers whose scopes have a common part in *S*. But again, we can show that proof procedures in quantification theory must make use of steps in which the number of individuals we are considering together in their relation to each other increases. Thus, at least some quantificational proofs are synthetic and so also, by extension, are some of the theorems proved.

On Hintikka's reconstruction, the paradigmatic synthetic method in quantification theory is the natural deduction rule of existential instantiation, that is, a rule that permits us to move from an existentially quantified sentence $(Ex)(Fx)$ to a sentence instantiating or specifying it, for example $F(a/x)$,

[20] See the *Critique of Pure Reason*, for example at A217/B264: "For through mere concepts of these things, analyse them as we may, we can never advance from one object and its existence to the existence of another or to its mode of existence." Interestingly, Kant makes this claim in a discussion of causality, possibly indicating at least one reason why he took causal statements to be invariably synthetic.

[21] Hintikka discusses certain aspects of this in the first section of "Kant Vindicated."

where *a* is a free individual symbol (or argument constant) and $F(a/x)$ the result of replacing x by a in F.[22] A condition on the soundness of this rule is that the instantiating symbol must be different from all the free individual symbols occurring earlier in the proof. Hence use of the rule inevitably introduces new representatives of individuals, what Hintikka takes Kant to mean by "intuitions," into the argument.[23] For this reason, mathematical proofs that can be reformulated as quantificational arguments in which the rule of existential instantiation is applied are synthetic.

Hintikka's view, as we mentioned earlier, is in many ways an extension of Beth's reconstruction.[24] In Beth's reconstruction, however, the emphasis falls in a slightly different place. Beth emphasizes the use of free variables in mathematical proofs to refer to *arbitrary* individuals in such a way that a generalized claim can be made about *all* individuals of a given type.[25] Thus in a typical proof, Euclid first introduces a par-

[22] I.e., the rule that allows us to move from talking about some x which is F to talking about a particular individual a which is F. See sections 9-12 of Hintikka's paper, "Quantifiers, Language-Games, and Transcendental Arguments," in *Logic, Language-Games and Information*.

[23] Hence there is no special connection, in the first instance, between the use of intuitions in mathematical arguments and Kant's claim, which the Transcendental Aesthetic seeks to establish, that intuitions (for us) are necessarily sensible.

[24] See footnote 4 for references.

[25] Thus echoing Locke, who says in the *Essay Concerning Human Understanding*, ed., A. C. Fraser (New York: Dover Books, 1959), Book IV, Chapter 1, pp. 173–174: "The immutability of the same relations between the same immutable things is now the idea that shows him, that if the three angles of a triangle were once equal to two right ones, they will always be equal to two right ones. And hence he comes to be certain, that what was once true in the case, is always true; what ideas once agreed will always agree; and consequently what he once knew to be true, he will always know to be true; as long as he can remember that he once knew it. Upon this ground it is, that particular demonstrations in mathematics afford general knowledge. If then the perception, that the same ideas will *eternally* have the same habitudes and relations, be not a sufficient ground of knowledge, there could be no knowledge of general propositions in mathematics; for no mathematical demonstration would be other than particular; and when a man had demonstrated any proposition concerning one triangle or circle, his knowledge would not reach beyond that particular diagram. If he would extend it further, he must renew

ticular individual, in a step known as *ecthesis*—"For let *ABC* be a triangle, I say that in the triangle *ABC* two sides taken together in any manner are greater than the remaining one, etc."—with respect to which the proof is carried out, and then draws a general conclusion—"Therefore, in any triangle, etc."[26] Kant seems to have just this procedure in mind at A716/B744: the geometer, in contrast to the philosopher, "at once begins by constructing a triangle." After carrying out an argument with respect to *this* construction, the geometer, "through a chain of inferences guided throughout by intuition . . . , arrives at a fully evident and universally valid solution of the problem." In the same way, proofs in algebra traditionally turn on reasoning carried out with respect to free variables. More generally, we can say that any proof, as cast into quantificational form, of a universal proposition that turns on the use of free variables in the argument is synthetic. What is important in such proofs, and what allows us to draw general conclusions, is that we assume nothing about the particular individual or singular term introduced except what is already given in the premises.[27] Such an individual is iden-

his demonstration in another instance, and so on; by which means one could never come to the knowledge of any general proposition."

[26] The notion of *ecthesis*, and a corresponding emphasis on free variable proofs in mathematics, is already present in Aristotle (see the *Prior Analytics*, 25, 15). Lukasiewicz points out the connection between this use of *ecthesis* and the rule of existential generalization in quantification theory, and the connection between the use of individual representatives of general concepts and sensibility. See his *Aristotle's Syllogistic from the Standpoint of Modern Formal Logic* (Oxford: Oxford University Press, 1951), pp. 60ff. "It is sense perception alone," Aristotle says in the *Posterior Analytics*, 81b7, "which is adequate for grasping the particulars." A suggestion Hintikka makes is that this tradition explains at least in part Kant's claim that all intuitions are sensible, or are "grasped" through sense perception; only sense perception provides knowledge of *individuals*.

[27] This is the point Kant seems to be making when he insists in the Preface to the second edition of the *Critique* that if the mathematician "is to know anything with *a priori* certainty he must not ascribe to the figure anything save what necessarily follows from what he has himself set into it in accordance with his concept" (Bxii).

tified by both Hintikka and Beth with an "intuition" in Kant's sense.[28]

The Beth-Hintikka reconstruction has two general features not yet mentioned. One is that it draws the line between logic and mathematics at monadic predicate logic.[29] According to Hintikka, Kant in fact draws the line just at this point. Hence, even if mathematics could be "reduced" to general quantification theory, it still would not cease being synthetic for Kant, since that part of logic too is synthetic.

The other aspect of the Beth-Hintikka reconstruction is that it places the emphasis on analytic *methods*; analytic sentences are simply those which can be shown to be true by strictly analytic methods. Analytic methods, recall, are those which do not introduce new individuals (or representatives thereof) into the discussion, and so on. Admittedly, there are reasons in Kant's text for concentrating on the character of the methods used to establish mathematical propositions.[30] But at the same time, there are passages that are difficult if not impossible to reconcile with this emphasis. To refer back to a passage already cited, Kant comments on his predecessors as follows at B14: "For as it was found that all mathematical inferences proceed in accordance with the principle of contradiction . . . it was supposed that the fundamental propositions of the science can themselves be known to be true through that principle. This is an erroneous view." It could not be stated more clearly that all mathematical inferences are analytic. The synthetic character of the propositions of mathematics is a function of some feature of the propositions

[28] It is worthy of note that Leibniz, Kant's dialectical opponent in the philosophy of mathematics, vigorously rejects proof procedures that turn on the use of *ecthesis*. See his comments on Locke's above-mentioned view in the *Nouveaux Essais*, Livre IV, Chapitre 1, #9.

[29] See Hintikka's paper, "Distributive Normal Forms in First-Order Logic," in *Logic, Language-Games and Information*, for some of the details and some of the complications in his development of this point.

[30] In addition to the passages from the Transcendental Doctrine of Method already mentioned, Kant suggests in the Introduction to the *Critique* that the *proof* of "7 + 5 = 12" requires intuition, and for this reason is synthetic.

themselves and not of the way in which they come to be established.[31]

the Lambert-Parsons reconstruction[32]

Recall that on the reductionist account of mathematics mathematical propositions are logical truths or substitution instances of logical truths by way of definitions, and so are analytic. Recall also that I characterized logical truths as sentences true in or of all possible worlds. One difficulty for the reductionist account is that there are true mathematical propositions—there exists an empty set, there are an infinite number of prime numbers, there exist at least two points, etc.—that are not true in every possible world, not, for example, in worlds in which the empty set does not exist or that do not contain at least two points.[33] Hence there are mathematical propositions that are, as Kant said, synthetic. Roughly classed, they are those propositions which make an existential claim.[34]

[31] See Lewis White Beck, *Studies in the Philosophy of Kant* (Indianapolis: The Bobbs-Merrill Company, Inc., 1965), pp. 89–90: "The real dispute between Kant and his critics is not whether the theorems are analytic in the sense of being strictly deducible, and not whether they should be called analytic now when it is admitted that they are deducible from definitions, but whether there are any primitive propositions which are synthetic and intuitive."

[32] This second reconstruction is a result in very large part of conversations with my former colleague Karel Lambert in connection with his work on free logic. See his paper, "On Logic and Existence," *Notre Dame Journal of Formal Logic*, VI (1965), pp. 135–141. Charles Parsons, in his very interesting paper "Kant's Philosophy of Arithmetic," places the same sort of emphasis on the importance of existence assumptions, although he links Kant's claim about the synthetic character of arithmetic with the sensibility of intuitions in a very different way. One untoward consequence of Parsons' view, at least on my interpretation, is that on it there is no distinction to be made between mathematics and logic.

[33] Remember that on our characterization of a "possible world," a world consisting of a single point is possible.

[34] Although Kant does not, to my knowledge, claim that every synthetic proposition is existential, he states clearly that every existential sentence is, *ipso facto*, synthetic. See the first *Critique*, A598/B626: ". . . if . . . , we admit, as every reasonable person must, that all existential propositions are syn-

Following Kant, let us take Euclidean geometry as our mathematical paradigm. An axiomatic formulation of Euclidean geometry, far from undermining Kant's view, allows one to see in a very sharp way the existential character of geometry. In Euclid's *Elements,* where the axioms are set down in the form of principles of construction, this existential character is obscured by the pictorial aspect of proof. The fundamental point, however, is that in carrying out constructions we are asserting the existence of mathematical individuals. And it is principally in virtue of this fact that mathematical proofs and propositions are synthetic. That all intuitions (in this case, mathematical individuals) are, for us human beings, necessarily sensible (a result of the Aesthetic and not a corollary of the definition of "intuition") explains how and in what sense the propositions of mathematics are *evident*, and hence supplies an additional reason for saying that they are synthetic. In the next chapter, I will develop this theme in more detail. Thus we have as a typical axiomatization (that suffices, in fact, for only a fragment of plane geometry):

undefined terms: point, line

G.1: Every line is a collection of points.

G.2: There exist at least two points.

G.3: If p and q are points, then there exists one and only one line containing p and q.

G.4: If l is a line, then there exists a point not on l.

---

thetic, how can we maintain that the predicate of existence can be rejected without contradiction? This is a feature which is found only in analytic propositions, and is indeed precisely what constitutes their analytic character." Kant does claim that every synthetic sentence presupposes an object; to this extent it has an existential presupposition. See the important letter to K. L. Reinhold of May 12, 1789: the principle of synthetic judgments is "unequivocally presented in the whole *Critique*, from the chapter on the schematism on, though not in a specific formula. It is this: *all synthetic judgments of theoretical cognition are possible only by the relating of a given concept to an intuition*." Translated by Arnulf Zweig in his edition of *Kant's Philosophical Correspondence, 1759–99* (Chicago: The University of Chicago Press, 1967), p. 141.

G.5: If l is a line, and p is a point on l, then there exists one and only one line containing p parallel to l.[35]

It is possible, furthermore, to draw a line where the making of existence claims begins. General quantification theory can be so reformulated as to contain no theorems that make an existence claim, more precisely, no theorems beginning with "there is a . . ." or "(Ex)" and whose scope is the entire formula.[36] The same point can be made by saying that the theorems of quantification theory, as reformulated à la Lambert, are true in *all* possible worlds, counting the empty domain as a possible world also. Hence mathematics, to speak rather generally, cannot be "reduced" to the reformulated theory. It *can* be "reduced" to set theory, but among the axioms of set theory are several that make an explicit existential claim.

We can put the matter in sharper focus by considering a particular axiomatization of set theory, for example, the Zermelo-Fraenkel version.[37] Zermelo-Fraenkel set theory comprises, in addition to the axioms of standard quantification theory, three different sorts of axioms. There are axioms that make no existence claim, *viz.* the axiom of extensionality; axioms that make a conditional existence claim ("given set A, there exists a set B such that . . ."), *viz.* the axioms of choice and regularity, among others; and axioms that are un-

[35] This "parallel postulate" is, in fact, equivalent to the proposition that Kant habitually uses to illustrate the synthetic character of geometry, that the sum of the interior angles of a triangle is equal to two right angles.

[36] One might say that for (a rationally reconstructed) Kant, "logic" is best understood as identical with universally free quantification theory. A quantification theory is "free" insofar as the rule of inference *Fa*/*(Ex)Fx* (existential generalization) is dropped, "universally free" insofar as no existential statement is provable. The first point is connected with a point made earlier that for Kant not all singular terms refer; the second has to do with the "empty" character of logical truths. See the papers by T. Hailperin, "Quantification Theory and Empty Individual Domains," *Journal of Symbolic Logic*, 18 (1953), pp. 197–200; K. Lambert, "Free Logic and the Concept of Existence," *Notre Dame Journal of Formal Logic*, 1967, pp. 133–144; and R. K. Meyer and K. Lambert, "Universally Free Logic and Standard Quantification Theory," *Journal of Symbolic Logic*, 33 (1968), pp. 8–26.

[37] This way of putting the point was suggested to me by Bas van Fraassen.

conditionally existential, *viz.* the null set axiom ("there exists a null set") and the axiom of infinity ("there exists a set containing at least all natural numbers," understood in the Zermelo way).[38] Assuming that the axiom of extensionality could be eliminated by way of a definition,[39] we could then say that set theory is quantification theory plus existence postulates.

This suggests a reason for drawing the line between logic and mathematics between logic without existence assumptions and mathematics (which rests on them), and a reason for saying that the latter cannot be "reduced" to the former.[40]

Hintikka has at least this much support for his reconstruction: the logic Kant explicitly describes in his lectures on logic resembles monadic predicate logic and is, in other respects, traditionally Aristotelian. But the argument Kant gives for the synthetic character of mathematical propositions and the

[38] See Quine, *Set Theory and Its Logic,* #39.

[39] To wit, "$S_x =_{df} (y)(y = z \leftrightarrow (z)(z \varepsilon y \leftrightarrow z \varepsilon x))$," in which case the quantifiers in the other axioms would have to be restricted to S (sets).

[40] My former colleague E. W. Kluge has brought to my attention a very late (1924/25) note of Frege's, "Neuer Versuch der Arithmetik," in the *Nachgelassene Schriften*, H. Hermes, F. Kambertel, and F. Kaulbach, ed. (Hamburg: Felix Meiner Verlag, 1969), pp. 298–302, in which he gives up the claim that arithmetic reduces to logic ("Ich habe die Meinung aufgeben müssen, dass die Arithmetik ein Zweig der Logik sei und dass demgemäss in der Arithmetik alles rein logisch bewiesen werden müsse"), on grounds somewhat like those I have just set out, and proposes instead to derive arithmetic from geometry, taking geometrical intuition ("die geometrisch Erkenntnisquelle") as the fundamental source of mathematical knowledge and truth.

To my knowledge, Russell never went so far as to admit that arithmetic was in any sense intuitive, but he was worried about the kind of existence claim that the axiom of infinity, in particular, made. In the second edition of *Principia Mathematica*, in fact, rather than assert it as an axiom, Russell and Whitehead include it as part of the hypothesis in the statement of each theorem in whose proof it figures; "on the assumption that there are an infinite number of individuals, then *T*" (where "*T*" is some particular theorem). An initial difficulty with this maneuver is that we can never detach the theorem in question, and hence never apply it to anything. Thus, on the hypothetical construal, the utility of much of mathematics is lost. For Kant the loss is greater, for, as we shall see, a proposition of mathematics is *meaningful* in his view only to the extent that it can be applied.

general philosophical position in which this argument is embedded suggest unmistakably that for him the crucial difference between mathematics and logic turns rather on the question of existential import.[41]

On Leibniz's view, mathematical existence is limited to logical existence: what is thinkable without contradiction. Mathematical proofs can be carried out with respect to any object whose description is consistent; for the purposes of our proof, we can assume the existence of such an object. For Kant, on the other hand, mathematical existence is limited not to "logical existence," but more narrowly to what can be *constructed.* To put it in a slightly different way, according to Kant one cannot prove the existence of anything on the basis of logical principles,[42] not even the existence of mathematical objects. This is, for example, the point made in a crucial passage at A60/B85 of the *Critique*: "But since the mere form of knowledge, however completely it may be in agreement with logical laws, is far from being sufficient to determine the material (objective) truth of knowledge, no one can venture with the help of logic alone to judge regarding objects, or to make any assertion. We must first, independently of logic, obtain reliable information. . . ."

The same point can be developed in the context of remarks made in the first chapter. Truths of logic hold in all possible worlds, truths of mathematics hold in just some possible worlds.[43] The possible worlds in which the truths of mathematics hold are not just sets of self-consistent concepts (to follow the "concept" characterization of "possible world" suggested earlier), but sets of concepts that are "constructible."

[41] It is a question, once again, of distinguishing between Kant's explicit, traditional views on logic and the logical perspective to which his position commits him.

[42] Hence the need to reformulate quantification theory so as to omit otherwise provable sentences like "$(Ex)(x = x)$" (i.e., "at least one thing exists").

[43] For Leibniz, by way of contrast, the truths of mathematics hold in all possible worlds. See the *Theodicy,* #351, where he asserts that the three-dimensional truths of Euclidean geometry hold in every possible world, from which it follows that Euclidean geometry is the only possible (consistent) geometry.

To use an expression already introduced, truths of mathematics hold in all "really possible" worlds. As Kant puts the point at B268, a passage to which we shall again refer:

"It is, indeed, a necessary logical condition that a concept of the possible must not contain any contradiction; but this is not by any means sufficient to determine the objective reality of the concept, that is, the possibility of such an object as is thought through the concept. Thus there is no contradiction in the concept of a figure which is enclosed within two straight lines, since the concepts of two straight lines and of their coming together contain no negation of a figure. The impossibility arises not from the concept in itself, but in connection with its construction in space, that is, from the conditions of space and of its determination. And since these contain *a priori* in themselves the form of experience in general, they have objective reality; that is, they apply to possible things."

Mathematical propositions describe the class of "really possible" worlds, and *a fortiori* the actual world. But there are logically possible worlds in which they do not hold. It is in virtue of just this fact that they are synthetic.

constructibility, pure intuitions, and objective reality

There are a number of ways in which Kant's metaphysics of experience can be viewed as a generalization of his philosophy of mathematics. Though later I will amplify and support this remark, here I want to say a little more about certain concepts, as they emerge in Kant's philosophy of mathematics, which prove important in the generalization.

Mathematics, in contrast to logic, is not "empty" or "merely formal." It has "content," and for this reason is synthetic rather than analytic. I have tried to indicate one way in which "having content" can be understood: the propositions of mathematics have existential import, whereas the propositions of logic do not. This fact has as a further consequence that the propositions of mathematics, unlike those of logic, do not hold in or of all possible worlds.

But the notion of "having content" has at least one more closely related feature. Mathematical propositions are *mean-*

ingful. I suggest that this comes to saying that at least they have truth values. On the face of it, this claim seems patently mistaken. Kant says, for example, that the truth of analytic propositions "can always be adequately known in accordance with the principle of contradiction"[44] and this would seem to guarantee that analytic as well as synthetic propositions have truth values and hence that they are equally meaningful. I think there is evidence, however, in addition to Kant's characterization of analytic propositions generally as "empty" and "merely formal," which suggests that analytic propositions do not necessarily have truth values.

There is, for example, a passage (which has often proved difficult to interpret) at B16 of the first *Critique*:[45]

"Some few fundamental propositions, presupposed by the geometricians, are, indeed, really analytic, and rest on the principle of contradiction. But, as identical principles they serve only as links in the chain of method and not as principles; for instance, $a = a$; the whole is equal to itself; or $(a + b > a)$, that is, the whole is greater than its part. And even these propositions, though they are valid according to pure concepts, are only admitted in mathematics because they can be exhibited in intuition."

The apparent, although here unstated, reason why analytic propositions must pass this requirement if they are to be admitted into mathematical arguments is that *all* propositions admitted into mathematical arguments must be "intuitive" in content. But, as I will argue in a moment, to have "intuitive" content is for the subject term of a proposition to refer to a *really possible* object. And, as I have suggested in the first chapter, only sentences whose subject terms are "really possible concepts" in this way have truth values. On this reading, identities flanked by singular terms that purport to refer to objects we are not capable of experiencing—"the round square = the round square" is a limiting case—are for that reason neither true nor false. They are simply meaningless.[46] To

[44] *Critique of Pure Reason*, A151/B190.

[45] Following Kemp Smith's rearrangement of the text.

[46] *Critique of Pure Reason*, A155/B194ff.

square this claim with the characterization of analytic propositions at A151/B190 quoted earlier, we could say that for Kant an analytic proposition if true is necessarily true, that is, if its subject term is a "really possible" concept then the truth of the proposition can be determined by inspection.[47]

This way of construing analytic propositions is supported, moreover, by a section of Kant's argument against the posibility of an ontological proof of the existence of God. In particular, Kant claims that:

"If, in an identical proposition, I reject the predicate while retaining the subject, contradiction results; and I therefore say that the former belongs necessarily to the latter. But if we reject subject and predicate alike, there is no contradiction; for nothing is then left that can be contradicted. To posit a triangle, and yet to reject its three angles, is self-contradictory; but there is no contradiction in rejecting the triangle together with its three angles" (A594/B622).

This is to say, I think, that if the existential presupposition is not satisfied, then the judgment does not have a truth value. Kant puts his point in a misleading way by saying that if you reject the subject you reject the contradiction, for he has already asserted that *any* judgment of the form $a = a$ is analytic, hence *any* judgment of the form $a \neq a$ is self-contradictory.[48] What he wants to say is that if you reject the subject, then the principle of contradiction cannot be used to establish the truth of the judgment. Thus, once a triangle is *posited*, it is true, analytically, that it have three angles. But if a triangle is not posited, then no judgment about "it" is either true or false.[49]

[47] Note that Kant says both that truth "is the agreement of knowledge with its object" (A58/B82) and that "An analytic proposition carries the understanding no further; for since it is concerned only with what is already thought in the concept, it leaves undecided whether this concept has in itself any relation to objects. . . . The understanding (in its analytic employment) is concerned only to know what lies in the concept; it is indifferent as to the object to which the concept may apply" (A259/B314).

[48] One way out of this difficulty would be to deny that all analytic propositions are *judgments*.

[49] It would not be an "objective" proposition.

That mathematical propositions are meaningful comes to saying also that they have *objective reality*. In the passage already referred to at A220/B268, Kant equates this important concept with what I have been calling "real possibility." In another passage, at A155/B194 of the *Critique*, objective reality is connected with meaningfulness:

"If knowledge is to have objective reality, that is, to relate to an object, and is to acquire meaning and significance in respect to it, the object must be capable of being in some manner given. Otherwise the concepts are empty; through them we have indeed thought, but in this thinking we have really known nothing; we have merely played with representations. That an object be given . . . means simply that the representation through which the object is thought relates to actual or possible experience. Even space and time, however free their concepts are from everything empirical, and however certain it is that they are represented in the mind completely *a priori*, would yet be without objective validity, senseless and meaningless, if their necessary application to the objects of experience were not established. . . . Apart from these objects of experience, they would be devoid of meaning. And so it is with concepts of every kind."

A number of different points are made in this passage. I want to discuss three of them. The first, already made, is that for a proposition to have a knowable truth value it must be objective, i.e., the object referred to by its subject term must be capable of being given. But in that case there are analytic propositions without knowable truth values. In formulating such propositions we are merely playing with representations.

The second point I want to select from the passage has to do with the connection Kant makes between objective reality and meaningfulness. A proposition is meaningful to the extent that objects corresponding to its subject term are capable of being experienced by us. But this claim in turn has two different aspects, developed in successive sections of the *Critique of Pure Reason*. In the Aesthetic (and in the Transcendental Doctrine of Method) a proposition is meaningful to the extent that it is "intuitive," that is, to the extent that its concepts can be "constructed." The propositions of mathematics are synthetic; they have content. In that case, however, mathematical propositions presuppose the possibility of corresponding intuitions. Since mathematics is at the same time undeniably *a priori*, the corresponding intuitions must themselves be *a priori*.[50] In the Analytic, on the other hand, a proposition has objective reality to the extent that it applies to the objects of experience. The truth of mathematical propositions can be shown through intuitive representation and demonstration. But their application to our experience requires a separate argument, which it is the task of the Analytic to supply. Thus, while the mathematician can "construct" his concepts, the full "objective reality" of mathematics rests, in the final analysis, on the philosopher's ability to show that experience is such that given propositions of mathematics (must) apply.[51] Kant's deep-rooted assumption is that objectivity (objective reality) requires objects, as the passage at A155/B194 makes clear, and it is not until the Analytic that the required objects are specified.

The third point I want to extract from the passage at A155/B194 has to do with my claim that the propositions of mathematics have existential import. The word "existence" is being used here in a special sense. We drew the line between logic and mathematics, between universally free quantifica-

[50] The first motive is mentioned to Reinhold in Kant's letter of May 12, 1789: "All synthetic judgments of a theoretical cognition are possible only by the relating of a given concept to an intuition. If the synthetic judgment is an experimental judgment, the intuition must be empirical; if the judgment is a priori synthetic, there must be a pure intuition to ground it" (in Zweig, *Kant's Philosophical Correspondence*, p. 141). The second motive is mentioned in another letter to Reinhold of May 19, 1789: "Mathematics is the most excellent paradigm for the synthetic use of reason, just because the intuitions with which mathematics confers objective reality upon its concepts are never lacking" (Zweig, p. 146).

[51] One of the referees has helped me to make this distinction clearer. In chapter 3, I discuss the general relation between the mathematician's and the philosopher's tasks; in chapter 4 I deal with the application of mathematics to the world of our experience.

tion theory and classical quantification theory-set theory, on the basis of the fact that the latter but not the former involves existence claims. But the line drawn between existence and non-existence is not between the actual and the possible, but between the "really possible" and the merely possible. To assert the existence of something from a mathematical point of view, then, is not to assert that it actually exists, but that it is a really possible object, one that could be experienced by us, the guarantee of which is its constructibility. It is with such remarks in mind that we are to read the following passage at A719/B747:

"It would therefore be quite futile for me to philosophize upon the triangle, that is, to think about it discursively. I should not be able to advance a single step beyond the mere definition, which was what I had to begin with. There is indeed a transcendental synthesis (framed) from concepts alone, a synthesis with which the philosopher alone is competent to deal; but it relates only to a thing in general, as defining the conditions under which the perception of it can belong to possible experience. But in mathematical problems there is no question of this, nor indeed of existence at all, but only of the properties of the objects themselves, [that is to say], solely in so far as these properties are connected with the concept of the objects."

Mathematical problems do not have to do with existence proper, but they do have to do with existence as given an epistemological twist, with "real possibility." Why Kant thought that a particular set of mathematical objects are "really possible" or why our experiencing capacities are limited in certain ways, is a question for the next chapter.

I referred early in this chapter to the Transcendental Doctrine of Method. It is of crucial importance in understanding Kant's philosophy of mathematics. In this section of the *Critique*, Kant draws an extended contrast between the "synthetic activities" of mathematicians and the "analytic activities" of philosophers. The contrast turns on his claim that mathematicians but not philosophers can "construct" their concepts. But in at least one respect the contrast is misleading.

For in Kant's view the philosopher is engaged ultimately in trying to guarantee the real possibility, objective reality, and meaningfulness of his own concepts as well as those of the mathematician. This involves him, on analogy with Kant's philosophy of mathematics, in something like the attempt to "construct" his concepts.[52] As we shall see, this comes, in the case of Kant's metaphysics of experience, to an attempt to "construct" the Categories, in the case of Kant's metaphysics of nature, to an attempt to "construct" the concept of matter.

[52] See the quotation from the *Metaphysical Foundations of Natural Science* used to preface this book. This is not Kant's technical sense of "construct," but something analogous to it. Philosophical concepts are "constructed" when their application to experience generally has been shown.

Chapter 3: geometry, Euclidean and non-Euclidean

KANT and Aristotle are, in my view, the two greatest western philosophers. They are also the only two philosophers, to my knowledge, whose views often seem to have been decisively refuted by developments in science. In the case of Aristotle, it is generally believed that the scientific revolution of the 16th and 17th centuries destroyed his position, not only in details, but also its central theme, the primacy of teleological explanation. In the case of Kant, it is no less widely held that his philosophy of mathematics was disproved by the logicist reduction of Frege and Russell, and his theory of geometry undermined, first, by the development of non-Euclidean geometry and next, by its application to the physical world in Einstein's theory of relativity. In contrast, no one seems to think that developments in science bear in the least on Plato's theory of ideas, or Berkeley's immaterialism, or Moore's ethical intuitionism, let alone the theories of Hegel or Heidegger.

As far as Kant's theory of geometry is concerned, the majority opinion is well expressed by Rudolf Carnap. In light of the development of non-Euclidean geometry, and the use made of it by Einstein:

"It is necessary to distinguish between pure or mathematical geometry and physical geometry. The statements of pure geometry hold logically, but they deal only with abstract structures and say nothing about physical space. Physical geometry describes the structure of physical space; it is part of physics. The validity of its statements is to be established empirically—as it has to be in any other part of physics—after rules for measuring the magnitudes involved, especially length, have been stated. In Kantian terminology, mathematical geometry holds indeed *a priori*, as Kant asserted, but only because it is analytic. Physical geometry is indeed synthetic; but it is based on experience and hence does not hold *a priori*. In neither of the two branches of sciences which are called 'geometry' do synthetic judgments *a priori* occur. Thus Kant's doctrine must be abandoned."[1]

This passage contains at least four different claims: that mathematical and physical geometry, and by implication mathematical and physical space, must be distinguished; that the propositions of mathematical geometry are analytic *a priori*; that the propositions of physical geometry are synthetic *a posteriori*; that the theory of geometry that the first three claims embody is entailed by or, at least, is consistent with (in a way in which Kant's position is not), the scientific developments since the publication of the *Critique of Pure Reason*. I will begin by discussing the first three claims, then respond to the fourth by trying to clarify the nature of Kant's commitment to Euclid, and finally make some brief comments on the Kantian view of space.

the analyticity of geometry

As Carnap puts it, "mathematical geometry holds indeed *a priori*, as Kant asserted, but only because it is analytic." How can the analytic character of geometry be shown? One suggestion, following one of Kant's characterizations of "analytic," is to say that a proposition of mathematical geometry is analytic insofar as its negation is self-contradictory or in conjunction with other such propositions leads to a contradiction. But this suggestion will not work. We can consistently negate given propositions of mathematical geometry, for example, the axiom of parallels. In fact, it was the unsuccessful attempt to show that negating the parallel axiom led to contradiction that resulted in the development of non-

[1] From the "Introductory Remarks" to the English edition of Hans Reichenbach's classic treatment of the subject, *The Philosophy of Space and Time*, translated by Maria Reichenbach and John Freund (New York: Dover Publications, Inc., 1958), p. vi. Even philosophers who otherwise disagree with a Carnap-like assessment of Kant's position concur. For example, P. F. Strawson: "There seems no doubt that these views are, at least to a very great extent, correct." *The Bounds of Sense* (London: Methuen & Co., Ltd., 1966), p. 278.

Euclidean geometries. For the essential feature of these geometries is that they consistently replace the parallel axiom with a negation of it.[2] Thus, if we operate with this concept of analyticity, the development of non-Euclidean geometries, far from undermining Kant's claim that geometry is synthetic, only serves to support it.

Sometimes geometrical propositions are said to be analytic insofar as they are deducible with the aid of definitions from logic. But, as we saw in the last chapter, if we take "logic" as universally free quantification theory, then this reduction cannot be carried out. Essentially the same point can be made in this way. If geometrical propositions were analytic in the sense that they followed from fundamental logical principles, and from definitions, they would be logical truths,[3] sentences true in or of all possible worlds. But the development of non-Euclidean geometries reveals that Euclidean geometry is not true in all possible worlds, hence not analytic in this sense either. There are possible worlds that these non-Euclidean describe, which is just to say that these geometries are consistent.[4] Still another, not quite equivalent, way of putting the same point is to say that under some interpretation of the non-logical constants that appear in the sentences of any (consistent) geometry, these sentences will come out true and under another interpretation they will come out false. But if these sentences were logical truths, then we should expect

[2] This is not quite right in the case of so-called Riemannian or elliptical geometry (since adjustments have to be made in some of the other axioms as well), but the present point is not affected.

[3] In the sense indicated earlier, in which, for example, "All bachelors are unmarried," counts as a logical truth.

[4] As already noted, Kant makes the point, presumably against Leibniz, at B268 of the *Critique*: "there is no contradiction in the concept of a figure which is enclosed within two straight lines, since the concepts of two straight lines and of their coming together contain no negation of a figure." It was Kant's appreciation of the fact that non-Euclidean geometries are consistent (possibly something of which his correspondent, the mathematician J. H. Lambert, made him aware) that, among several different considerations, led him to say that Euclidean geometry is synthetic. The further development of non-Euclidean geometries only confirms his view. See Frege, *The Foundations of Arithmetic*, p. 20.

that they would come out true under all interpretations of the non-logical constants that appear in them.

In brief, the propositions of mathematical geometry cannot be said to be "analytic" in any of the senses we have so far attributed to Kant. But this is only the beginning of the story.

Kant follows a tradition dating from Euclid that divides the fundamental principles of geometry into two groups, *axioms* ("common notions") and *postulates*. We set down a list of postulates for Euclidean geometry in the last chapter. Among the axioms are:

A.1: Things that are equal to the same thing are also equal to each other.

A.2: If equals are added to equals, the wholes are equal.

A.3: If equals are subtracted from equals, the remainders are equal.

A.4: Things that coincide with one another are equal to one another.

A.5: The whole is greater than the part.

The fact that both are assumed from the outset in carrying out proofs and the belief that the postulates are only disguised logical truths has led to a blurring of the distinction between axioms and postulates, and it is now common to use "axiom" and "postulate" interchangeably. But it is just this distinction on which much of Kant's claim for the synthetic character of geometry turns.

In fact, there are two ways in which the distinction between axioms and postulates has been made traditionally.[5] One (Aristotelian) way may be summarized as follows:

"Every demonstrative science must start from indemonstrable first principles; otherwise, the steps of a demonstration would be endless. Of these indemonstrable principles, some are (a) common to all sciences, others are (b) particular, or

[5] See the 5th-century mathematician Proclus, *A Commentary on the First Book of Euclid's Elements*, translated with introduction and notes by Glenn Morrow (Princeton: Princeton University Press, 1970), p. 143. Proclus lists three ways of making the distinction. I have concentrated on two, following Proclus' own discussion.

peculiar to the particular science; . . . the common principles are *axioms*, most commonly illustrated by the axiom that, if equals be subtracted from equals, the remainders are equal. In (b) we have the *genus* or subject-matter, the *existence* of which must be assumed."[6]

The contrast here is between the axioms as principles of logic common to all the sciences and the postulates as principles defining a particular subject-matter. Thus, in the development of non-Euclidean geometries, the axioms remain the same while the postulates, the fifth in particular, are changed. The truth of the axioms is independent of the particular subject-matter under discussion, while the truth of the postulates is not.

The other (more properly Euclidean) way of marking the contrast is between the axioms as non-constructive, the postulates as constructive principles.[7] "Constructive" is to be taken quite literally. As Euclid sets them out, the postulates are principles that assert the possibility of drawing particular geometrical figures.

P.1: To draw a straight line from any point to any point.
P.2: To produce a finite straight line continuously in a straight line.
P.3: To describe a circle with any center and distance.

But "constructive" can be taken in a less literal sense as well, as suggested in the last chapter. For the various principles of construction can be reformulated as sentences that make an existential claim. Thus, "To draw a straight line from any point to any point" goes over as "For any two distinct points, there exists exactly one line containing them." The postulates formulate the existence assumptions that a given set of geometrical propositions involve. In a geometri-

[6] T. L. Heath, *The Thirteen Books of Euclid's Elements* (Cambridge: Cambridge University Press, 1908), I, p. 119.

[7] As Proclus points out in his *Commentary*, these two ways of making the distinction are not quite equivalent. On the first, all of the principles of Euclidean geometry listed in chapter 2 are postulates; on the second, only the first three are postulates.

cal proof, we first "postulate" a subject matter and then proceed to reason about it.[8]

I have introduced these brief remarks on the notion of a "postulate" not only to reinforce the argument of the last chapter, and the emphasis there laid on the connection between the existential and synthetic character of mathematical propositions, but to indicate a leading motive for Kant's insistence that geometrical proofs essentially involve constructions. These remarks should also help set the stage for the argument on which the positivist theory of geometry espoused by Carnap seems to depend.

Following Hilbert, we are to think of mathematical geometries as "uninterpreted" sets of sentences. The method is as follows. Take the axioms—correctly speaking, the postulates—that Euclid set down: between any two points a straight line can be drawn, any finite straight line can be extended in a straight line, etc. Now replace the non-logical expressions "point," "line," etc., by schematic letters. Thus "disinterpreted," the first postulate becomes

> For any two distinct *P*'s, there is an *S* to which each of them bears the relation *B*,

and the second postulate (slightly reformulated) becomes

> For any *S* such that there are two distinct *P*'s each bearing the relation *B* to it, there is another *S* to which each of these *P*'s bears the relation *B* but to which only one of them bears the relation *E*.

We could interpret the *P*'s, *S*'s, *B*'s, *E*'s as points, lines, etc. But, in that case, the geometry would no longer be a purely mathematical one, but rather an applied geometry.[9] And we

[8] See the *Critique of Pure Reason,* A234/B287: "in mathematics a postulate means the practical proposition which contains nothing save the synthesis through which we first give ourselves an object and generate its concept—for instance, with a given line, to describe a circle on a plane from a given point. Such a proposition cannot be proved, since the procedure which it demands is exactly that through which we first generate the concept of such a figure."

[9] Insofar as these expressions are not "interpreted," they are often said to be "implicitly defined" by the postulates of the theory. Thus the entire mean-

could interpret the *P*'s, *S*'s, *B*'s, and *E*'s in a variety of other ways as well, under some of which interpretations they would be true and under others false. But from a mathematical or abstract point of view, we are interested in none of these "interpretations." Our interest is solely in the *relations* that these sentences assert between the schematic letters.[10]

Now if we take mathematical geometry to consist of sets of uninterpreted sentences, then there is a rather short argument for its analytic character. Let us say that a sentence is analytic just in case it is "empty," or has no "content," or cannot be refuted by empirical information. Then the sentences of a mathematical geometry are analytic. For not only do they not express propositions that can be refuted, they do not express propositions at all. As Carl Hempel put it in a well-known article: "We see therefore that indeed no specific meaning has to be attached to the primitive terms of an axiomatized theory, and in a precise logical presentation of axiomatized geometry the primitive concepts are accordingly treated as so-called logical variables. . . . For this very reason, the postulates themselves do not make any assertion which could possibly be called true or false!"[11]

Many philosophers seem to think that this is the end of the matter. The case for the analytic, hence *a priori*, character of

---

ing of "*P*," for example, derives from the syntactic role it plays in the postulates, and not from the images, etc., we commonly associate with the notion of a "point." In the same way, it is often said that a mathematical geometry is a "formal theory," with no meaning at all attached to its axioms.

[10] As the late 19-century geometer Pasch, who did so much to advance this conception of mathematical geometry, put it: "Indeed, if geometry is to be deductive, the deduction must everywhere be independent of the *meaning* of geometrical concepts, just as it must be independent of the diagrams; only the *relations* specified in the propositions and definitions may legitimately be taken into account. During the deduction it is useful and legitimate, but in *no* way necessary to think of the meanings of the terms; in fact, if it is necessary to do so, the inadequacy of the proof is made manifest." Quoted by R. L. Wilder, *The Foundations of Mathematics,* second edition (New York: John Wiley and Sons, Inc., 1965), pp. 7–8.

[11] "Geometry and Empirical Science," in H. Feigl and W. Sellars, ed., *Readings in Philosophical Analysis* (New York: Appleton-Century-Crofts, 1949), p. 244.

the "propositions" (with quotation-marks now supplied) of mathematical geometry has been proved. There is a mistake here, however, the mistake of confusing the making of a distinction with the giving of an argument.[12] If one makes a distinction in the way indicated, between "interpreted" and "uninterpreted" systems of geometry, and further identifies mathematical geometry with the latter and physical geometry with the former, then one has a reason for saying that mathematical geometry is, perhaps in a somewhat special sense, analytic and physical geometry is synthetic. But until it is shown that one must or should make this distinction, and assume the concept of analyticity with which it is aligned, not just that one can, the case against Kant has been begged.

There are at least three arguments hovering in the background that might turn the trick. The first, already rejected because of a false premise, is that mathematical geometry is a branch of logic, the sentences of which depend only on the logical constants they contain for their truth and falsity. The second argument is that the sentences of geometry can be taken in only two ways: either (a) as "interpreted," in which case they receive a *physical* interpretation and so are subject to confirmation or disconfirmation via empirical experiments, or (b) as "uninterpreted." Since we do not believe that the sentences of a mathematical geometry are subject to confirmation or disconfirmation (they are either "assumptions" or deductive consequences of same), we must hold that they are "uninterpreted," hence in the sense indicated by Hempel "analytic." The third argument contends that the only way we can construe the sentences of a mathematical geometry consistent with the developments since Euclid is as "uninterpreted." To quote Hempel once again:

"The fact that these different types of geometry have been developed in modern mathematics shows clearly that mathematics cannot be said to assert the truth of any particular set of geometrical postulates; all that pure mathematics is interested in, and all that it can be said to establish, is the deductive con-

[12] Karel Lambert has taught me the importance of making *this* distinction.

sequences of given sets of postulates and thus the necessary truth of the ensuing theorems relatively to the postulates under consideration."[13]

Kant does not consider the possibility of a mathematical geometry as a system of "uninterpreted" statement-forms. That he does not consider this possibility does not, of course, entail that his position is outdated or mistaken. But he does distinguish explicitly between applied mathematics and pure mathematics, "the very concept of which implies that it does not contain empirical, but only pure *a priori* knowledge."[14] If Kant's position rests on conflating pure and applied geometry, it is not that he does not make the distinction but that he fails to appreciate it; or perhaps that he understands it in a somewhat different way, because for Kant the propositions of (pure) Euclidean geometry are both meaningful and true.

Kant's distinction between pure and applied geometry is between *a priori* and empirical knowledge. Strawson[15] is tempted to construe this distinction not as one between "interpreted" and "uninterpreted" sets of sentences, but as one between "interpretations" of different kinds. We can, according to Strawson, distinguish between "physical" and "phenomenal" interpretations. A physical interpretation defines "lines," for example (or the schematized "*L*'s" of the disinterpreted postulates), as the paths of light rays in a homogeneous medium, and "points" and "planes" in terms of other physical conditions. A "phenomenal" interpretation, on the other hand, identifies the various geometrical entities with the spatial "looks" of things. On the "phenomenal" interpretation, straight lines are "just the looks themselves which physical things have when, and in so far as, they look straight."[16] Strawson intends the notion of a "phenomenal" interpretation to capture what Kant means when he talks about the construction of a geometrical concept in pure intuition.

[13] *Readings in Philosophical Analysis*, p. 243.

[14] *Critique of Pure Reason*, B15.

[15] *The Bounds of Sense*, part v.

[16] *Ibid.*, p. 282.

The "phenomenal" interpretation has a double-barrelled use. On the one hand, it implies that the positivist alternative—give the sentences either a "physical" interpretation or none at all—is not inevitable. We can give them an interpretation that is "non-physical," yet that is at the same time connected with sense-given spatial objects in such a way that the sentences so interpreted are synthetic. On the other hand, Strawson's claim that the postulates of Euclidean geometry are true "solely in virtue of the meanings they contain, but these meanings are essentially phenomenal, visual meanings,"[17] and his accompanying contention that the only figures we can visualize are Euclidean, gives a sense to and supports Kant's thesis that Euclidean geometry is *a priori* as well.

Strawson's interpretation has at least this much going for it, that Kant too intends the concept of pure intuition to provide backing *both* for the thesis that the sentences of Euclidean geometry are synthetic *and* for the thesis that they are *a priori*. Kant seems to have thought that since geometrical proofs required constructions (and could not be carried out simply by an analysis of concepts), the propositions proved must be synthetic, and that since the constructions could be carried out in the imagination (independent of experience) these same propositions were known *a priori*. But the thesis that they are synthetic can be supported without appealing to pure intuition or a phenomenal interpretation and Strawson's defense of the *a priori* thesis is porous.

The defense is porous simply because mathematical geometry does not depend on the construction of figures or the exhibition of "visual meanings." They are dispensable. Moreover, if the propositions of a phenomenal geometry are true, they would appear to be no more than contingently true, for it is a contingent matter that we can visualize in one or another geometrical manner.[18] Finally, whatever the actual geometrical properties of and relations between physical ob-

[17] *Ibid.*, p. 283.

[18] These points are made by James Hopkins, "Visual Geometry," *The Philosophical Review*, LXXXII (1973), pp. 3–34.

jects, and whatever our judgments or perceptions of same, the actual geometrical properties of and relations between visual objects (objects in our visual field) are *not* in fact Euclidean but Riemannian.[19]

At the same time, it is possible to understand a "non-physical" interpretation of mathematical geometry somewhat this side of the doctrine of "visual meanings" and to support the thesis that even a "pure" geometry is synthetic without recourse to the mistaken belief that constructions are essential. For Euclid, Aristotle, and, I have suggested, for Kant, a mathematical geometry has a subject matter that is *postulated.*[20] To say what the subject matter of a set of propositions is is to indicate the class of possible worlds of or in which these propositions would be true.

This can be made formally more precise.[21] Following Patrick Suppes,[22] we say that to axiomatize a theory is to define a particular set-theoretical predicate. Thus, to axiomatize Euclidean or Riemannian or Lobachevskyan geometry is to define "a Euclidean space," etc., as an "*n*-tuple $S = P, L, \ldots$ where P is a non-empty set (the points of S), L is a

[19] See R. B. Angell, "The Geometry of Visibles," *Nous,* VIII (1974), pp. 87–117. As Angell points out, the thesis that the geometry that correctly describes the actual configurations of the visual field is a two-dimensional, elliptical geometry was first advanced by Thomas Reid, *An Inquiry into the Human Mind on the Principles of Common Sense*, sixth edition, esp. Ch. IV, Section 9, "The Geometry of Visibles." Angell, correctly I think, traces Strawson's position to a confusion (in his talk about the "looks things have") between *judgments* about the geometrical properties of physical objects and the geometrical properties of the objects of the visual field.

[20] The notion of "postulation" is important to Kant's general philosophical position, although I make very little of it here. From one point of view, the doctrine of the synthetic *a priori* derives from it: propositions are synthetic insofar as they have a subject matter, but *a priori* insofar as that subject matter has been postulated. That the subject matter has been "postulated" *by us*, moreover, is the reason why we can gain certain knowledge of it, a claim already made by Hobbes with regard to both political science and geometry (see *De Corpore*, III, 9). As Kant puts it in the Preface to the second edition of the *Critique*, "we can know *a priori* of things only what we have ourselves put into them."

[21] Again, I am indebted to Bas van Fraassen for the suggestion, although he is not to be held responsible for its elaboration.

[22] In his *Introduction to Logic* (Princeton: Van Nostrand Reinhold, 1957), chapter 12.

family of subsets of P (the lines of S), etc., such that for any two members x,y of P there is a unique member l of L containing both x and y," etc. (listing the other postulates, all of which now become part of the set-theoretical definition). It should be clear that on such an axiomatization the geometrical postulates have, as such, no "factual" content, and to this extent Carnap, Hempel, and those who share their views are correct. But at the same time, it should be clear that a geometrical theory so axiomatized has a subject matter, *viz.*, a family of set-theoretical structures.[23]

Now although these points tell to some extent against the positivist claim that mathematical geometry does not have a "subject matter," but is concerned exclusively with the syntactic relations obtaining between the various geometrical terms, they do not deal directly with the associated claim that the postulates of a mathematical geometry *implicitly define* the basic notions of the particular geometry involved.[24] To put it in terms of our set-theoretical characterization, the "meaning" of the various basic notions is given by the set-theoretical structures that satisfy particular lists of postulates. It might be said, following Hilbert, that this is just what a formally rigorous axiomatization shows. But there are at least three objections to this view,[25] each of which bears on features of Kant's theory of geometry.

In the first place, it is hard to see in what sense an "implicit

[23] And in *this* respect is not to be distinguished from similar axiomatizations of, say, particle mechanics.

[24] See Ernest Nagel, *The Structure of Science* (New York: Harcourt, Brace & World, Inc., 1961), p. 91: ". . . insofar as the basic theoretical terms are only implicitly defined by the postulates of the theory, the postulates assert nothing, since they are statement-forms rather than statements (that is, they are expressions having the form of statements without being statements), and can be explored only with the view of deriving from them other statement-forms in conformity with the rules of logical deduction."

[25] See Frege's penetrating critique of Hilbert's "implicit definition" view of geometry in his papers "On the Foundations of Geometry." An excellent English translation of them by E. W. Kluge has recently appeared, *Gottlob Frege on the Foundations of Geometry and Formal Arithmetic* (New Haven: Yale University Press, 1971). See also Paul Bernays' review, "Max Steck, Ein unbekannter Brief von Gottlob Frege," *The Journal of Symbolic Logic*, 7 (1942), pp. 92–93, and E. W. Beth, *The Foundations of Mathematics*, pp. 344ff., 490ff., and 514ff.

definition" is a *definition*. Although we might want to say that a basic notion is implicitly defined by the other basic notions with respect to the postulates of the theory insofar as the one basic notion is provably definable in terms of the others,[26] it is part of the view we are considering that *all* the basic notions are implicitly defined.[27] In the second place, the postulates of Euclidean geometry, for example, do not define the basic notions—point, line, plane, etc.—even in the wider sense of "define." For, as observed earlier, in the wider sense they "define" a certain family of set-theoretical structures, or models, that satisfy these postulates. But many of these structures or models do not have any particular geometrical (that is, spatial) characteristics;[28] yet it is surely part of the meaning of "point," "line," and "plane" that they are spatial entities, quite apart from any visual images or principles of construction we might associate with them. In the third place, and perhaps most damaging for those for whom the last point is question-begging, the various set-theoretical structures that a particular list of postulates "define" are not isomorphic; that is, they differ in cardinality. To put it in an equivalent way, the postulates for at least some of these geometries are not absolutely categorical.[29] It would seem to follow from these three objections that the postulates of a theory do not "de-

[26] Indeed, such an implicit definition could be turned into an explicit definition satisfying the standard criteria of eliminability and non-creativity.

[27] As Frege says, it is like solving an equation in which all the terms are unknown.

[28] In fact, the postulates for geometry listed earlier are satisfied by *any* collection of objects (chairs, numbers, people, *inter alia*) where "point" means any element and "line" means any pair of elements of the collection. The only further requirement is that the collection have at least four elements.

[29] In particular, geometries that have the power and means of expression of Tarski's $E_{2'}$. See Tarski's paper, "What is Elementary Geometry?" in L. Henkin, P. Suppes, and A. Tarski, eds., *The Axiomatic Method* (Amsterdam: North-Holland Publishing Co., 1959). Depending on the precise formulation of "elementary geometry" chosen, of course, different metamathematical results can be proved. Of the formalized geometries Tarski discusses, $E_{2'}$, seems to me best to approach Kant's informal conception. It should be added that the formalization of $E_{2'}$, is first order and that there are second-order axiomatizations that are (for a given dimension) categorical.

fine," even implicitly, the basic objects of the theory. In addition to the partial characterization afforded by the postulates, a geometrical theory must also appeal to intuition if the characterization of its notions or objects is to be complete.[30] At this point, there is no need to add that this appeal to intuition inevitably yields Euclidean results. For we are so far concerned only with the synthetic, and not also the *a priori*, character of geometry. Without putting much weight on Kant's doctrine of pure intuition, I have tried to show that the positivist arguments on behalf of the analyticity of geometry fail. It is enough with regard to the final argument, in particular, to show that there must be a way of further restricting the family of set-theoretical structures or models that the postulates of a mathematical theory "implicitly define" before we can characterize the subject matter of such a geometry adequately.

geometry and space

As noted earlier, Kant does not consider the possibility of construing a mathematical geometry as a system of "uninterpreted" axioms. He does not have a conception of geometry except as a theory about real or imagined space. This fact tends to blur the distinction between the thesis that geometry is synthetic and the thesis that it is *a priori*, and it requires considerable reshaping of his position to deal directly with the attacks on the first thesis alone. At the same time, it is precisely as a theory about real space that the positivists locate their main disagreements with Kant. For, on their view, the recent history of physics has decisively refuted Kant's claim that Euclidean geometry is *a priori*.

Kant wants to say that Euclidean geometry is *a priori* not only in the sense that it describes a set of possible worlds, but that it is the geometry that describes the set of "really possible" worlds, that is, worlds that we are capable of experiencing, and *a fortiori* the actual world. On this reading of "*a*

[30] See Beth, *The Foundations of Mathematics,* p. 643.

[31] Perhaps the most difficult thing about Kant's theory of geometry is distinguishing between the premises and the conclusions.

priori," Euclidean geometry is necessarily descriptive of our experience.

Kant does have an argument for the priority of geometry in this sense, or rather a related set of arguments, that involves the concept of *a priori* intuitions. The argument does not establish all that it has been traditionally taken to establish, but it indicates the kinds of pressures Kant thought he was under and the kinds of problems his theory was designed to solve.

One part of the argument was sketched in the last chapter. It begins with the premise[31] that the propositions of Euclidean geometry are "meaningful"; they have truth values. But if a proposition has a truth value, then its corresponding "existential presupposition" must be satisfied. That is, an object corresponding to the subject term of the proposition must be capable of being given, the proposition must describe a "really possible" state of affairs. The only guarantee of "real possibility" in the case of pure concepts such as those of a mathematical geometry is an *a priori* "construction"; and the only concepts capable of being "constructed" are those of Euclidean geometry. We can call this the "existence problem," understanding "existence" in the epistemological way, as "real possibility." The doctrine of *a priori* intuitions is intended as a solution to this problem. The possibility of providing *a priori* intuitions for Euclidean geometry guarantees satisfaction of its presuppositions.[32]

[32] Kant makes this clear, I think, in the course of his reply to Eberhard (in the section entitled, "Concerning the objective reality of those concepts to which no corresponding sensible intuition can be given, according to Mr. Eberhard"). Eberhard had maintained that "the mathematicians themselves (have) completed the delineation of entire sciences without saying a single word about the reality of their object." To which Kant replies by commenting on the exemplary procedures of Apollonious. "Apollonious first constructs the concept of a cone, i.e., he exhibits it *a priori* in intuition (this is the first operation by means of which the geometer presents in advance the objective reality of his concept). He cuts it according to a certain rule, e.g., parallel with a side of the triangle which cuts the base of the cone . . . at right angles by its summit, and establishes a priori in intuition the attributes of the curved line produced by this cut on the surface of the cone. Thus, he extracts a concept of the relation in which its ordinates stand to the parameter, which concept, in this case, the parabola, is thereby given *a priori* in intuition. Con-

The second part of the argument has to do with the application of mathematics to the physical world. How, to put the question in a Kantian way, is such an application possible? What guarantees a *fit* in this case between our geometric concepts and the world? This might be called "the application problem." It is, I think, the problem around which Kant's views on geometry, and more generally his theory of science, revolve. Kant's solution is to say that the "construction" of geometric concepts (the exhibition of corresponding *a priori* intuitions) is governed by the same conditions that govern our perception of physical objects. Kant sometimes puts the claim this way: any "really possible" world has a particular spatial-temporal form and it is this form that the *a priori* intuitions we are capable of generating represent. A pure intuition is an intuition of the form of a really possible object. Thus, the existence assumptions in Euclidean geometry are justified because of the structure or form of sensibility that in some sense "conditions" physical objects, the existence of which in general we come to know through perception.[33] The same activity that constructs concepts "constructs" objects, and thereby guarantees a fit between them.

Presumably Kant saw the "existence problem" and the "application problem" as two sides of the same coin.[34] They come together in Kant's claim that the propositions of Euclid-

---

sequently, the objective reality of this concept, i.e., the possibility of the existence of a thing with these properties, can be proven in no other way *than by providing the corresponding intuition*." *On a Discovery According to which Any New Critique of Pure Reason Has Been Made Superfluous by an Earlier One* (1790), translated and with an introduction by Henry E. Allison, *The Kant-Eberhard Controversy* (Baltimore: The Johns Hopkins University Press, 1973). The passage quoted is on p. 110.

[33] See Hintikka, "Kant on the Mathematical Method," sections 17–19.

[34] See the *Prolegomena*, #8: "Concepts are indeed of such a nature that we can very well make some of them for ourselves *a priori*, without ourselves standing in any immediate relation to the object; namely the concepts that only contain the thought of an object in general, e.g., the concepts of quantity, cause, etc. But even these, to provide them with meaning and sense, still require a certain use *in concreto*, i.e., application to some intuition through which an object of these concepts is given to us."

ean geometry have "objective reality." On the one hand, to say that a proposition has objective reality is to say that it has a (knowable) truth value. On the other hand, to say that a proposition has objective reality is to say that it applies to objects. But how, on the one hand, do we guarantee truth values, and how, on the other hand, do we guarantee application to objects? The answer, once again, is to claim that Euclidean geometry is *a priori*.

Now it is important to notice that this argument establishes at most that if a geometry has objective reality then it must be *a priori*. But the argument does not show that Euclidean geometry has objective reality. This is a separate thesis. It is consistently maintained by Kant, but it figures as a premise of his arguments, not as a conclusion. Presumably Kant thinks that Euclidean geometry provides a uniquely correct description of space. It is also true that the doctrine of *a priori* intuitions is in part intended as an explanation of how we come to know the geometric truths we do know. But it is not for Kant a properly *philosophical* task to show that Euclidean geometry does provide a uniquely correct description of space or that we are capable of visualizing only in a Euclidean manner.[35]

[35] The comments of one of the referees have helped me to see this more clearly. The point is emphasized by Peter Mittelstaedt in the course of an interesting discussion of Kant's theory of geometry in *Philosophical Problems of Modern Physics* (Dordrecht: D. Reidel Publishing Company, 1976). As he says (pp. 48–49): "Although three-dimensionality, metricability, and the relations of space as a whole are constitutive of experience, this is not the case for Euclidicity. The latter is, in fact, generated by the construction of the concepts relevant to Euclidean geometry. That Euclidean geometry . . . is applicable to experience, is occasioned by the fact that the same geometrical constructions can be executed in intuition as in empirical space. This is demonstrated in the principle of the axioms of intuition. This principle not only ensures the validity of Euclidean geometry in experience, but also the empirical validity of geometry in general. . . . Experience is therefore possible before one sets forth a scientific geometry that structures this experience in a particular way." and adds as a note (p. 84): ". . . one must . . . bear in mind the difference . . . between mathematical and transcendental-philosophical propositions. Whereas the principles of pure reason follow from the conditions of the possibility of experience, the truth of mathematical propositions is based on the construction of its concepts and on immediate intuition."

The philosophical task is to show what the presuppositions of objective reality are, in this case to show "how geometry is possible."[36] Mathematicians have already demonstrated, on the basis of mathematical (including, for Kant, the construction of figures) not philosophical considerations, which geometric propositions are true. It is the philosopher's job to show that the geometric concepts involved are not mere "fictions" but must be descriptive of space. It follows that it is not by itself damaging to Kant's philosophical position to discover that space, in the large, is not Euclidean,[37] although he would have to modify his explanation of how we come to know the geometric truths that we do. What *is* damaging to Kant's position, and one of the cornerstones of the positivist view, is that it is an apparent consequence of the theory of relativity that it is a contingent fact that Euclidean geometry does or does not appropriately describe space, to be decided on the basis of (admittedly very sophisticated and theory-laden) observations and measurements. But if there is something wrong with Kant's thesis concerning the *a priori* character of geometry, and if the thesis is intended as a solution to a problem, then, it might be suggested, perhaps there is something wrong with the problem.[38]

The problem from Kant's point of view stems, I think,

[36] Note that Kant's argument in the Transcendental Exposition of the Concept of Space in the *Critique* and in the corresponding sections of the *Prolegomena* is entirely general and simply assumes the existence of a body of geometrical truth. The form of the argument does not require that space be Euclidean.

[37] Whatever this might mean. Perhaps it would be better to say that it is not appropriately described in Euclidean terms.

[38] Some commentators contend that in fact Kant does not have a problem, that, given a particular concept of "mobility" (the mobility of rigid objects of finite extension), various geometries can be constructed as *a priori* theories (*a priori* valid of the objects of experience). Thus, for example, Mittelstaedt, *Philosophical Problems of Modern Physics*, p. 81: "On the basis of unrestricted mobility, it is possible to construct Euclidean geometry as an *a priori* theory, which is necessarily valid for all objects of the prescientific domain of experience. Similarly, Riemannian geometry can be established on the basis of restricted mobility. It too is an *a priori* theory in the sense explained, and is necessarily valid for all objects of experience."

from what I have called his "anti-reductionism." In connection with his theory of geometry it has two aspects. One concerns the synthetic *a priori* character of mathematics. If mathematics were analytic, then there would be no "application problem," for as disguised logical truths the propositions of mathematics would apply universally.[39] But Kant thinks, for the reasons indicated, that the propositions of mathematics are synthetic. If mathematics were *a posteriori*, then there would presumably be no "application problem" either, for in that case the propositions of mathematics would be no more than inductive generalizations on our experience of the world. But Kant thinks, in company with all but Mill, that the propositions of mathematics are *a priori*, here in the limited sense of "*a priori*," which means something like "not borrowed (or derived) from experience." It is the fact that he takes geometry to be both synthetic and *a priori* that confronts him with a problem, to which he apparently thinks the doctrine of *a priori* intuitions is the only solution.[40]

The other aspect of Kant's "anti-reductionism" that contributes to his theory of geometry is his insistence that talk about physical objects cannot be "reduced" in the idealist way. On his view, emphasized in the Refutation of Idealism, physical objects have an important kind of autonomy. But the very autonomy of physical objects creates, he seems to think, a problem about the application of mathematics to them. How is it that we can be guaranteed, as apparently we are in physics, that the objects of the natural world will be

[39] I.e., there is no "application problem" for logic. This has to be qualified in the light of the passage at B16 to which I have already drawn attention: propositions of the form "$a = a$," for example, have objective reality only to the extent that the singular terms and bound variables that they contain refer to really possible objects.

[40] It is this same problem, generalized, that he takes to be the fundamental problem of metaphysics: how are synthetic *a priori* judgments possible? I.e., how can we guarantee that judgments reducible neither to truths of logic (or definitions) nor to sequences of sense experiences have application to the world of our experience? If the problem is generalized, so too is the solution: the world of our experience, at least as regards its form, is in some sense our "construction."

mathematizable, if they are given to us *a posteriori*, in perception? The solution, once again, is that the guarantee is possible only by way of the assumption that *we* contribute the form of such objects.[41]

space and spatiality

One thing that very much complicates Kant's discussion of geometry and space in the Transcendental Aesthetic is his failure there to distinguish as sharply as he does elsewhere between spatiality and space. When he tries to establish on the basis of the Metaphysical Exposition of the Concept of Space that space is an *a priori* concept, he has spatiality, what he elsewhere calls "space in general,"[42] in mind. The Transcendental Exposition of the Concept of Space, on the other hand, has to do not with space in general but with determinate space.

In trying to establish in the Metaphysical Exposition that space is an *a priori* concept, Kant argues, against Leibniz, that the concept of space cannot be "reduced." His argument is

[41] Note I to #13 of the *Prolegomena* makes this perfectly clear. As Kant there says in part: "It will always remain a remarkable phenomenon in the history of philosophy that there was a time when even mathematicians who were also philosophers began to doubt, not indeed of the correctness of their geometrical propositions insofar as they merely concern space, but the objective reality and application to nature of this concept and of all geometrical determinations of it. They were anxious whether a line in nature might not consist of physical points and true space in the object, of simple parts, although the space that the geometer thinks about can in no way consist of these. They did not recognize that it is this space in thought which itself makes possible physical space, i.e. the extension of matter; that it is not a quality of things in themselves but only a form of our faculty of sensible representation; . . . and that space, as the geometer thinks it, being precisely the form of sensible intuition which we find in ourselves *a priori* and which contains the ground of the possibility of all outer appearances (as to their form), it must agree necessarily and in the most precise way with the propositions of the geometer, which he draws from no fictitious concept, but from the subjective foundation of all outer appearances, namely sensibility itself. In this and no other way can the geometer be secured as to the undoubted objective reality of his propositions against all the chicaneries of a shallow metaphysics. . . ."

[42] *Metaphysical Foundations of Natural Science*, 484.

that Leibniz's reductive account of space is circular: individuation of objects, on which Leibniz's construction of space depends, already presupposes that bodies have individuating spatial positions.[43] Commentators have often pointed out that Kant uses the existence of incongruent counterparts in the early essay "Concerning the Ultimate Foundation of the Differentiation of Regions in Space," the *Prolegomena*, and the *Metaphysical Foundations of Natural Science* to argue very different conclusions. What is infrequently recognized is that all these arguments turn on the same consideration: that the existence of incongruent counterparts shows that the principle of the identity of indiscernibles cannot be used in a non-circular way to individuate objects and hence to construct space in terms of them, for the only differences that incongruent counterparts display are *spatial* differences. It is in this sense that space is an *a priori* concept. The individuation of objects depends on their occupying spatial positions; therefore the concept of space cannot be "reduced" to an order of objects. All this argument requires is that objects be spatial in some very general sense. We can call it the spatiality thesis. It is independent of any further considerations regarding the nature of space.

In the Transcendental Exposition of the Concept of Space, however, Kant argues (for the reasons given in the last section of this chapter) that space must have a determinate, Euclidean character. This is a thesis not about the spatiality of objects but about the nature of space.

It is important to distinguish these theses for at least two reasons. First, the irreducibility of the concept of space is used by Kant to establish that space is "subjective" in the sense that our experience of objects presupposes it.[44] But Euclidean geometry is not "subjective" in this same sense, for its *a priori* status does not derive from the fact that the objects of our ex-

[43] See the parallel remarks on the construction of time in the note added to the second edition of the *Critique*, B53ff. When Kant insists that on his view time is *real* what he is insisting on in part is that it cannot be "reduced."

[44] What is presupposed by experience cannot be derived from it; therefore, Kant concludes, it must be "contributed by us."

perience are inevitably spatial. Running these two theses together results in the misleading suggestion that Euclidean geometry describes the form of our perceptual constitution.

The second reason for distinguishing the two theses is that unless we do we cannot make much sense out of many of the sections of the *Critique* that follow the Aesthetic. For in these sections Kant appears to extend his discussion of space and time. In particular, he argues that unless the Axioms of Intuition, the Anticipations of Perception, and the Analogies of Experience are presupposed, space and (especially) time will not have a determinate structure. If we take the claim that space and time have a determinate structure to have already been settled in the Aesthetic, then these sections seem superfluous at best. If, on the other hand, we construe the Aesthetic as establishing only the spatiality thesis, then the task of the following sections becomes clear. On this reading, the Transcendental Exposition of the Concept of Space[45] simply presupposes that space has a determinate structure in order to show that our *a priori* knowledge of such structure, via geometry, can only be explained on the thesis that space is "ideal."

[45] Added at the end of the discussion, so to speak, in the second edition of the *Critique*.

Chapter 4: the axioms of intuition

THE section of the *Critique of Pure Reason* called the "Axioms of Intuition" has received comparatively little commentary, and what it has received is unsatisfactory. On the one hand, there has been confusion about its relation to the Aesthetic. Commentators seem to argue either that it is a trivial consequence of the Aesthetic[1] or that it is incompatible with it.[2] On the other hand, many find it difficult to say exactly what the *argument* is,[3] or to relate it to any central theme of the Analytic,[4] sometimes even assigning it to a demand of the "architectonic."[5]

In this chapter I will sketch a reconstruction of Kant's remarks on the "axioms of intuition," try to relate them consistently to the Aesthetic, and bring them into the main line of the argument of the Analytic. The materials involved allow one to build a bridge from Kant's philosophy of mathematics to his philosophy of physics, and I will begin discussing the

[1] H. W. Cassirer: the assertion that all appearances are extensive magnitudes "does not more than make explicit something already implied by the assertion that space and time have universal application with regard to the objects of experience." *Kant's First Critique* (London: George Allen & Unwin, 1954), p. 136.

[2] According to Garnett, Kant's views on space in the Aesthetic and in the Analytic are "inconsistent." *The Kantian Philosophy of Space* (New York: Columbia University Press, 1939), p. 232.

[3] A. C. Ewing: "Kant holds both this and the next principle to have immediate certainty, so presumably he must be understood to be not so much proving as expounding and clarifying them." *A Short Commentary on Kant's Critique of Pure Reason* (Chicago: University of Chicago Press, 1930), p. 148.

[4] R. P. Wolff: ". . . the Axioms fall outside the chain of argument from consciousness to causation." *Kant's Theory of Mental Activity*, p. 238. Also P. F. Strawson: the "connexion with the general themes of the Analytic is tenuous and is made, as far as it is made at all, through the concept of 'synthesis.' " *The Bounds of Sense*, p. 31.

[5] N. Kemp Smith: ". . . the argument must have left Kant with some feeling of dissatisfaction. Loyalty to his architectonic scheme prevents such doubt and disquietude from finding further expression." *A Commentary to Kant's "Critique of Pure Reason,"* p. 347.

Metaphysical Foundations of Natural Science. In making the transition, which amounts to a transition from pure to applied mathematics, I hope also to illuminate the general structure of Kant's argument for the existence of synthetic *a priori* propositions.

extensive magnitudes

The principle of the Axioms is "All intuitions are extensive magnitudes," or, in the language of the first edition, "All appearances are, in their intuition, extensive magnitudes." This principle, Kant tells us in #24 of the *Prolegomena*, "subsumes all appearances, as intuitions in space and time, under the concept of *quantity*, and is thus a principle of the application of mathematics." What is the connection between the extensiveness of intuitions and the applicability of mathematics to them?

Let us begin with the concept of an extensive magnitude. For Kant, it is the concept of a magnitude "the representation of (whose) parts makes possible, and necessarily precedes, the representation of the whole" (A162/B203). The reference to "parts" suggests, as does the expression "extensive magnitude" itself, that spatial extension is at stake. But since Kant wants to say that temporal as well as spatial magnitudes are extensive, this cannot be what he has in mind. Rather, what Kant is driving at is that extensive magnitudes are *additive*; the "parts" referred to are simply units—of mass, of length, of time, for instance—the sum of each of which goes to make up a particular magnitude.

Before supporting this interpretation, let me make it more precise. The formula "All intuitions are extensive magnitudes" obscures the fact that it is properties, not objects or intuitions *per se*, that are extensive magnitudes.[6] An object has

[6] Clearly the formula makes no sense at all if we use our suggested reading of "intuition," *viz.*, "singular term." What has to be said is that the individuals to which our singular terms ostensibly refer are (or have) extensive magnitudes. Kant's claim is that singular terms refer just in case the individuals ostensibly referred to have a definite position in space and time. The Axioms of Intuition spell out one aspect of what it is to have a definite spatial-temporal position.

extensive magnitude, for example, with respect to its length, but not with respect to its color. Those properties are extensive magnitudes which are additive; i.e., there exists an empirical operation for them formally similar to addition in arithmetic. Thus, lengths can be added. The total length of two rods put end to end is the sum of their respective lengths. In the same way, "wholes" that have "parts" are extensive magnitudes in the sense that the "parts" may be added together to form these "wholes."

There is, moreover, an ancient tradition, dating from Aristotle ("quantity is that which has parts external to one another"), that identifies the notion of an extensive magnitude characterized in terms of additivity with the notion of a quantity. As Pierre Duhem puts it, in commenting on this tradition: "The essential character of any attribute belonging to the category of quantity is therefore the following: Each state of a quantity's magnitude may always be formed through addition by means of other smaller states of the same quantity; each quantity is the union through a commutative and associative operation of quantities smaller than the first but of the same kind as it is, and they are parts of it."[7]

If we assume that Kant belongs to this same tradition, much of his argument becomes clearer. Reconstructed from the first edition of the *Critique*, it would go as follows. All intuitions are, of necessity, sensible, hence have, again of necessity, spatial and temporal properties (results we can assume have already been reached in the Aesthetic). But spatial and temporal properties are additive; that is, they are quantities or extensive magnitudes. Therefore, intuitions (with respect to their spatial and temporal properties; presumably they have no others) are extensive magnitudes. Kant appears to give this proof of it: "I cannot represent to myself a line, however small, without drawing it in thought, that is, generating from a point all its parts one after another. Only in this way can the intuition be obtained. Similarly with all times, however small" (A162-3).

[7] *The Aim and Structure of Physical Theory,* translated by Philip Wiener (Princeton: Princeton University Press, 1954), p. 110.

I take it, however, that this autobiographical comment is not so much a proof of the premise as an illustration of it. If a "proof" is needed, it would probably go like this. All sensible intuitions, or perceptible objects, are spread out in space and time. A geometric point is not a perceptible object. But an object thus spread out has a shape and a size. And while the shape is not an additive property,[8] size is. Hence all sensible intuitions are additive with respect to at least some of their spatial and temporal properties. Admittedly the "proof" limps a little bit in the case of time. Why not suppose that objects exist for a temporally unextended instant? Kant's response is that time can be represented only spatially, e.g., as a line, in which case we can again carry through the above "proof." But this response requires the argument of the first and second Analogies of Experience to give it substance. If, in any case, there is *any* property an object has that is additive, we can call the object an extensive magnitude. And in the case of sensible intuitions it is apparent that they do have at least one such property (perhaps this is why Ewing is led to say that the principle is "self-evident").

This preamble enables us to explore with greater depth the connection between the extensiveness of intuitions and the applicability of mathematics to them. We have to ask ourselves under what conditions objects can be measured, realizing that measurement is not of objects *per se* but of properties. Under what conditions can numbers be assigned to them? The standard answer goes as follows. In the first place, objects can be measured when they can be arranged in some order that is isomorphic to the structure of some numerical system. In particular, if objects can be ordered by a transitive and asymmetric relation, then numbers can be assigned to them. Once objects can be ordered in this way, with respect to some property, they can be compared numerically. They can be measured.

Measurement has at least this "existential presupposition"; it assumes that the appropriate asymmetric and transitive rela-

[8] Hence not all spatial, and in this sense "extensive," properties are extensive, i.e., additive, properties.

tions hold. This claim involves two others: that these relations are *a priori* insofar as presupposed and synthetic insofar as they are not derivable from fundamental logical principles or merely definitional of "longer than." Thus, suppose two "axioms" for measurement:

$M.1$: $(x)(y)(Lxy \rightarrow \text{not-}Lyx)$,
$M.2$: $(x)(y)(z)(Lxy \,\&\, Lyz \rightarrow Lxz)$,

where the variables are to take physical objects as values and "Lxy" is to be read as "x is longer than y." These "axioms" are not simply inductive generalizations on our past experience with objects, nor are they in any profound sense "analytic." In this respect, like all constitutive principles, they stand somewhere in a middle ground between physics and logic. There are merely possible, but not "really possible," worlds in which there are three objects a, b, c such that Lab, Lbc, and Lca. Such objects would not be, in the appropriate sense, *rigid*. For the same reason, measurement as we ordinarily understand it could not be carried on.

Now although satisfaction of these topological conditions allows us to assign numbers to objects, hence to measure them, it does not allow us to do this in a precise way. We can say that one object is longer than another ("longer than" is an appropriate asymmetric and transitive relation), and assign a higher number to the former than to the latter. But satisfaction of these conditions does not allow us to say precisely *how long* either object is. A second set of metrical conditions must be satisfied before the question "how much?" can be answered. In particular, objects must be additive as well as orderable with respect to a given property before physical relations precisely correspond to numerical relations. Thus the application of mathematics to intuitions depends on their being extensive magnitudes. This is what Kant undertakes to show in the Axioms. As he summarizes the point at A165/B206: "This transcendental principle of the mathematics of appearances greatly enlarges our *a priori* knowledge. For it alone can make pure mathematics, in its complete precision, applicable to objects of experience."

One reason for attributing this view to Kant is that it makes sense of his argument, and relates the notions of quantity (pure concept), number (category), and extensive magnitude (principle) in an extremely natural way, far removed from the demands of the "architectonic." There are also a number of texts that support it. The view is explicit as early as the *Inaugural Dissertation*—"And it is not open to us to make intelligible even the *quantity* of space itself unless we should express that space by a number, after it has been related to a measure taken as a unity"[9]—and as late as the *Critique of Judgment*, #26. Further, Kant repeatedly emphasizes the fact that in the section entitled "Axioms of Intuition" no *axioms* of intuitions—i.e., axioms for arithmetic or geometry—are set out. Rather, what is set out is a condition (among others) that must be satisfied if such axioms are to have application. "In the Analytic," Kant says near the end of the *Critique*, "I have indeed introduced some axioms of intuition into the table of principles of pure understanding; but the principle there applied is not itself an axiom, but serves only to specify the principle of the possibility of axioms in general, and is itself no more than a principle derived from concepts. For the possibility of mathematics must itself be demonstrated in transcendental philosophy" (A733/B761).

the metric of space and time

As I have presented it, the argument of the Axioms as developed in the first edition of the *Critique* depends in a non-trivial way on the Aesthetic for one of its premises. So it is not, as is sometimes suggested, inconsistent with the Aesthetic. The charge of inconsistency rests largely on the apparent contrast between those passages in the Axioms in which Kant asserts that magnitudes are extensive "when the representation of the parts makes possible, and therefore necessarily precedes, the

[9] #15, corollary. From *Kant: Selected Pre-Critical Writings*, translated and with an introduction by G. B. Kerferd and D. E. Walford (Manchester: Manchester University Press, 1968). We might paraphrase this: unless the measurement axioms are taken to hold, not much *sense* can be made of the notion of the length of a physical object.

representation of the whole" (A162/B203) and the third argument of the Metaphysical Exposition of the Concept of Space where we are told that: "Space is not a discursive or, as we say, general concept of relations of things in general, but a pure intuition. For, in the first place, we can represent to ourselves only one space; and if we speak of diverse spaces, we mean thereby only parts of one and the same unique space. Secondly, these parts cannot precede the one all-embracing space, as being, as it were, constituents out of which it can be composed; on the contrary, they can be thought only as *in* it. Space is essentially one; the manifold in it, and therefore the general concept of spaces, depends solely on (the introduction of) limitations" (A24-25). The essential points of contrast are, first, between the notion of space as being generated and being given, and, second, between the notion of the parts of space preceding the whole and the whole preceding the parts.

There have been a variety of attempts to rescue Kant at these points. Many of them, following Ernst Cassirer, assert that in the Aesthetic Kant suppresses mention of the role of the understanding to better concentrate on the contributions of sensibility. But there is more at stake here than an expository strategy. The concept of space developed in the Aesthetic is amorphous in the sense that it has no particular metrical properties. Space is given, the argument of the Aesthetic goes, insofar as all our outer intuitions are inevitably spatial. It is one insofar as particular spaces are to be construed as parts of the same spatial framework in which all outer intuitions are "located." But *precise* location of these intuitions in the spatial framework is not yet possible. The space that is given and single has as yet no metric, and in this respect is not yet measurable.[10]

[10] Kant's discussion indicates that "given" space, the space of the Aesthetic, is a three-dimensional topological space. In *Thoughts on the True Estimation of Living Forces*, he had suggested that the topology of space is a function of the forces that bodies exert on each other, that the three-dimensionality of space in particular is implied by Newton's inverse square law. (Friedrich Überweg was later to provide a detailed proof of the connection in his *System der Logik,* 1882.) I know of no other attempt by Kant to give dimensionality a physical basis; if one takes the inverse square law as *a*

The contrast between the space and time of the Aesthetic and the space and time of the Axioms is made explicit by Kant in terms of a notion of *determinateness.* After asserting that magnitudes are extensive "when the representation of the parts makes possible, and therefore necessarily precedes, the representation of the whole," he adds, "I cannot represent to myself a line, however small, without drawing it in thought, that is, generating from a point all its parts one after another. Only in this way can the intuition be obtained. Similarly with all times, however small. In these I think to myself only that successive advance from one moment to another, *whereby through the parts of time and their addition a determinate time-magnitude is generated*" (A163-3/B203, my italics).

The force of this passage, I suggest, is that additivity of temporal properties makes a *determinate* time magnitude possible. The principle of the Axioms, in guaranteeing the application of mathematics "in its complete precision," at the same time makes possible determinate, measurable spatial and temporal magnitudes. But since space and time as given are (in the sense already indicated) metrically amorphous, an "act of synthesis"[11] is required if space and time, and intuitions generally, are to satisfy the appropriate metrical conditions. This point is repeated by Kant in his statement of the argument of the Axioms in the second edition of the *Critique*: ". . . appearances are all without exception *magnitudes*, indeed *extensive magnitudes.* As intuitions in space or time, they must

---

posteriori, and Kant did, then it would seem to follow that the dimensionality of space is merely contingent, and this thesis the "mature" Kant wanted to deny. Thus, although he might have used "incongruent counterparts" to provide a physical basis for dimensionality, in his later work Kant uses it to establish the "intuitive" character of space as well as the more general proposition "that *space in general* does not belong to the properties or relations of things in themselves" (*Metaphysical Foundations of Natural Science*, 484, my italics). For more discussion of some of these points, see Bas van Fraassen, *An Introduction to the Philosophy of Space and Time* (New York: Random House, 1970), pp. 134–138. Of course, the argument of the Axioms of Intuition requires that "given" space be metrizable as well.

[11] By which, roughly, we can understand the activity of conceptualizing space and time in certain ways.

be represented *through the same synthesis whereby space and time in general are determined*" (B203, last italics mine). In going beyond the general notions of spatiality and temporality developed in the Aesthetic to the concept of a determinate (measurable or additive) space and time in the Axioms, Kant is clearly not contradicting himself. It is also clear that the Axioms are not simply a trivial consequence of the Aesthetic.

More light can be thrown on these remarks by considering them in the historical context in which Kant worked out his conception of space and time. It is commonplace to point out that in the *Critique of Pure Reason* Kant breaks with the full Newtonian conception of space and time, in part to reconcile it with Leibniz's view. One feature of this break especially concerns us here. For Newton, space (and also time) has an intrinsic metric. It is worth quoting his view at length:

". . . the common people conceive these quantities (i.e., time, space, place and motion) under no other notions but from the relation they bear to sensible objects. And thence arise certain prejudices, for the removing of which it will be convenient to distinguish them into absolute and relative, true and apparent, mathematical and common . . . because the parts of space cannot be seen, or distinguished from one another by our senses, therefore in their stead we use sensible measures of them. For from the positions and distances of things from any body considered as immovable, we define all places; and then with respect to such places, we estimate all motions, considering bodies as transferred from some of those places into others. And so, instead of absolute places and motions, we use relative ones; and that without any inconvenience in common affairs; but in philosophical disquisitions, we ought to abstract from our senses, and consider things in themselves, distinct from what are only sensible measures of them. For it may be that there is no body really at rest, to which the places and motions of others may be referred . . . those . . . defile the purity of mathematical and philosophical truths, who confound real quantities with their relations and sensible measures."[12]

[12] *Principia*, Cajori edition, pp. 6, 8, 11. The selection follows Adolf Grün-

Clarke echoes the same position in his correspondence with Leibniz: "That it is not a bare relation of one thing to another, arising from their situation or order among themselves, is no less apparent, because space is a quantity, which relations (such as situation and order) are not."[13] In requiring that an act of "synthesis" be presupposed if time and space are to be measurable, and *a fortiori* all objects with respect to their spatial-temporal properties, Kant rejects the Newtonian view. For Kant, space and time do not have an intrinsic metric. Rather, a metric is "brought to" space and time by the understanding—that is, they are conceptualized in certain ways. The activity of the understanding thus makes measurement of objects with respect to their spatial and temporal properties possible. To repeat a point already made, the understanding thereby makes possible *determinate* space and time.[14]

My claim that, for Kant, space and time do not have an intrinsic metric is apt to be misleading against the background of recent discussion of the issues involved.[15] To say that space and time do not have an intrinsic metric means, generally, that it is only relative to "sensible measures," e.g., rods and clocks, that we can attribute a metric to them. But there are several different ways in which one might argue for this claim

---

baum, *Philosophical Problems of Space and Time*, second edition (Dordrecht: D. Reidel Publishing Company, 1973), pp. 4–5.

[13] Fifth Letter, #46, note. From the Alexander edition.

[14] See the *Prolegomena,* #38: "Space is something so homogeneous and in respect of all particular properties so indeterminate that there is certainly no hoard of natural laws to be found in it. . . . The mere universal form of intuition that is called space is indeed the substratum of all intuitions which can be designated to particular objects, and admittedly there lies in space the condition of the possibility and variety of intuitions. But the unity of objects is determined solely by the understanding, and according to conditions which lie in the nature of the understanding."

[15] See Grünbaum, *Philosophical Problems of Space and Time*, second edition, chapter 16 for an extended survey and Paul Horwich, "Grünbaum on the Metric of Space and Time," *British Journal for the Philosophy of Science*, 26 (1975), pp. 199–211, for a subsequent critique of Grünbaum's position. In what follows, I intend to leave it an open question whether Grünbaum's thesis is correct.

and, as a result, several different implications that one might draw from it. Reichenbach, for example,[16] argues that if we change our congruence standard (e.g., in the case of time, some process taken to be uniformly periodic), the metric changes. Since different congruence standards, and the alternative metrics that result from them, fit the same set of observed facts (i.e., there is no way of verifying that one or another of these standards is correct), Reichenbach concludes that our choice of a particular congruence standard must be conventional. Grünbaum, to take another example, rejects Reichenbach's argument. Alternative metrizability is not a sufficient condition for the conventionality of any particular metric for, Grünbaum maintains, there is a clear sense in which *discrete* spaces and times have an intrinsic metric despite the fact that they may be metrized in different ways. His argument develops that advanced by Riemann in his celebrated dissertation, "On the Hypotheses Which Lie at the Foundations of Geometry" (1854). It depends on the premise that space and time are not discrete "manifolds" (i.e., made up of countable spatial and temporal parts). If space or time were discrete, then their metric would be intrinsic, for in that case we could count the parts of which spatial or the "moments" of which temporal intervals were composed and assign magnitudes to them. In such a case we can say, with Riemann, that "the principle of metric relations is implicit in the notion of this manifold." If, on the other hand, space and time are continuous, as classical physics conceives them, then no such counting procedure is possible and the metric has to be "introduced."[17] It is in this sense that space and time are, for Grünbaum, "metrically amorphous," a fact that, he thinks, implies that the metric of continuous manifolds is "conventional."

Now, although Kant keeps insisting on the continuity, by

[16] *The Philosophy of Space and Time*, #8.

[17] To put it very briefly, the continuous sets of the qualitatively identical "points" that comprise the various spatial and temporal intervals are of equal cardinality and the sets have no other properties that might be used to distinguish the intervals.

which both he and Riemann seem to mean the density, of space and time,[18] there is no hint of anything like Riemann's argument in any of his writings nor would he accept Grünbaum's contention that by convention any number of geometries can be provided for the world. That synthesis is required for the measurability of space and time does not imply that the metric is a matter of choice. Along the same lines, and for reasons advanced in the last chapter, Kant would deny Reichenbach's contention that there exist alternate metrics for any "really possible" world and thus that the adoption of a particular congruence standard is conventional. Kant uses a different sort of epistemological argument to reach his conclusion, and it has very different implications. In brief, it involves turning Newton's position around. We cannot, and this is the fundamental premise, consider things in themselves, apart from sensible measures of them. But if our knowledge of the metric is relative to such "sensible measures," then we cannot claim that the metric is intrinsic. The notion of an intrinsic metric, thus understood, is without empirical significance.[19]

Kant's argument for the fundamentally extrinsic character of the metric of space[20] can be amplified. It is apparent from the passage quoted that for Newton there is a close connection between the notion of an intrinsic metric and the absolute theory of space.[21] On the absolute theory of space, there is

[18] See A169/B211: "the property of magnitudes by which no part of them is the smallest possible, that is, by which no part is simple, is called their continuity. Space and time are *continua quanta*. . . ."

[19] In his observation on the antithesis of the First Antimony (A431/B459), Kant makes explicit the application of this general principle to the measurement of space. After agreeing with the Leibnizian school concerning its attack on the "impossible assumption" of an absolute space and time, he adds: "Things, as appearances, determine space, that is, of all its possible predicates of magnitude and relation they determine this or that particular one to belong to the real. Space, on the other hand, viewed as a self-subsistent something, is nothing real in itself; and cannot, therefore, determine the magnitude or shape of real things."

[20] In the sense attributed to Kant. Presumably the same remarks would apply to time as well.

[21] That Kant assumes the same close connection is evident from his early

ultimately one set of co-ordinates that define the positions of all objects, quite apart from whatever "sensible measures" we may have of them. Clarke argues in the same vein: on Leibniz's relational theory of space no account can be given of spatial magnitude; the suggestion is that, since "space *is* quantity," such magnitude is intrinsic. But Kant, at least insofar as his "Critical" views are concerned, vigorously rejects the Newtonian theory of absolute space. By its very characterization, absolute space is not a possible object of experience,[22] it is not a "really possible" object. It is instead, as we shall see in a moment, a regulative idea. Hence, absolute space is not an object about which we can gain knowledge. The concept is, although thinkable (i.e., consistent), ultimately "empty" (i.e., without objective reality). The underlying assumptions here are, of course, those mentioned in the first chapter. Kant's mature view is that space is relational, a possible object of experience just insofar as perceptible, i.e., conceived in terms of those "sensible measures" we possess. This view had the consequence—a consequence requiring additional, highly controversial premises—that space is subjective and in some sense "ideal." Another, related consequence is the claim that the metric of space, or for space, is "contributed by us."

---

essay, "Concerning the ultimate foundation of the differentiation of regions of space" (1768), onwards. In that essay, he claims that on the basis of the existence of incongruent counterparts "it is clear that the determinations of space are not consequences of the situations of the parts of matter relative to each other; rather are the latter consequences of the former. It is also clear that in the constitution of bodies differences, and real differences at that, can be found; and these differences are connected purely with *absolute and original space,* for it is only through it that the relation of physical things is possible." Kerferd and Walford, *Kant: Selected Pre-Critical Writings*, p. 43. Kerferd and Walford point out that by the time of the *Prolegomena*, Kant, again on the basis of incongruent counterparts, had abandoned his commitment to the Newtonian concept of absolute space.

[22] For one thing, any possible object of experience occupies a spatial position, but clearly absolute space can occupy no such position. For another thing, absolute space is infinite, but infinite objects are not possible objects of experience.

motion and space

Kant's most extensive discussion of the notion of absolute space comes in his *Metaphysical Foundations of Natural Science*, published in 1786. The book is divided into four chapters, each corresponding, according to Kant's "architectonic" arrangement, to one of the four headings in the Table of Judgments. The "correspondence" is loose at best, just as it is between the Table of Judgments and the Categories in the *Critique of Pure Reason*. But, again as in the *Critique*, there is some point to it.

The first chapter is intended to correspond to the Axioms of Intuition. It is entitled "Metaphysical Foundations of Phoronomy" and has to do with what Kant calls the "pure doctrine of the quantity of motion."[23] While the Axioms are taken by Kant to insure the applicability of mathematics to objects in general, the apparent purpose of the Phoronomy is to show how such application can be guaranteed more specifically for bodies considered as movable points. This in turn comes down to showing how to "construct" a composition of motions, that is, to represent motions in such a way that they can be added, so that magnitudes may be assigned them.

Since the Phoronomy has to do with the pure doctrine of the quantity of motion, independent of any dynamical considerations, it is concerned only with the composition or addition of the motions of a point determined with respect to its direction and velocity. Kant's view, to put it in more modern terms, is that such motions are to be represented as spatially constructible vector velocities. In this way the assignment of magnitudes to motions and the consequent application of mathematical operations to them, a first step in the development of a mathematical physics, can be assured.[24]

[23] By which expression we can understand something like "pure kinematics."

[24] Although in this case the "sum" of the vectors is given by the parallelogram rule (which thus provides for the "spatial addition" of velocities) and not by their arithmetic sum.

What interests us here is not so much Kant's proof that composition of motions can be represented only in a particular way[25] as the remarks he makes concerning absolute space. Kant begins the Phoronomy with a defense of what he calls "relative space" and hence also of "relative motion" (since the motion of an object is defined as the change of its external relations to a given space). Insofar as space, or a space, is to be a possible object of experience, i.e., perceptible, it must be capable of being located with respect to some other space in which its movement may be perceived.[26] All perceptible space thus is relative to some other space in which we assume it to be located. The ideal limit to the procedure of "embedding" spaces in still more comprehensive spaces is "absolute space." But the concept of absolute space refers only to this ideal limit. It is a regulative idea, the principle of the construction of ever-larger spaces, and not an object of experience. "To make this absolute space an actual thing means to mistake the logical universality of any space, with which I can compare each empirical space as being included in it, for a physical universality of actual compass, and to misunderstand reason in its idea."[27] But since absolute space is "impossible" (i.e., although its concept is consistent, it is not applicable to an object of possible experience), so too is absolute motion.

[25] For a detailed discussion of the proof, see Vuillemin, *Physique et métaphysique kantiennes*, pp. 56–83, and Robert Palter, "Kant's Formulation of the Laws of Motion," *Synthèse*, 24 (1972), pp. 96–116.

[26] Without extensive apology, Kant takes motion to be the fundamental characteristic of matter. Matter as the movable is, in fact, the subject of the *Metaphysical Foundations of Natural Science*. At times, he suggests a quasi-physiological reason for taking motion to be fundamental: motion is required for "outer sense" to be affected, hence for perception to take place. Perhaps Kant is here simply taking a Cartesian premise for granted, that the distinction between matter and empty space depends on motion.

[27] Academy edition, IV, p. 482. This and all subsequent quotations of the *Metaphysical Foundations of Natural Science* are from the translation by James Ellington (Indianapolis: The Bobbs-Merrill Company, Inc., 1970). I will continue to refer to the pagination of the Academy edition; it is indicated in the margins of the Ellington translation.

All motion (construed "phoronomically" as velocity) is relative. This has an important consequence:

". . . in all relative motion the space itself, because it is assumed to be material, may be represented as at rest or as moved. The first occurs when, beyond the space with reference to which I regard a body as moved, there is no more extended space that includes this space (as when in the cabin of a ship I see a ball moved on a table). The second occurs when outside this space there is another space that includes this one (as, in the case mentioned, the bank of the river), since with regard to the riverbank I can view the nearest space (the cabin) as moved and the body itself (the ball) as at rest. Now, respecting an empirically given space, however extended it may be, it is utterly impossible to determine whether or not this space is itself moved with reference to a still greater space enclosing it. Hence for all experience and for every inference from experience, it must be all the same whether I want to consider a body as moved, or else consider the body as at rest and the space as moved, or else consider the body as at rest and the space as moved in the opposite direction with an equal velocity."[28]

The two concepts are equivalent. What Kant seems to have done, again on the basis of an epistemological argument, is to have inferred the correctness of Galileo's "principle of relativity": that with respect to a body in (uniform, rectilinear) motion, it is indifferent whether we treat it as in motion or at rest.

The case gives us, in fact, a concrete illustration of the presupposition relation we constructed in the first chapter. Consider a body moving at constant velocity. Is the body in fact in motion or is it at rest? On the presupposition of the existence of absolute space, we could assign truth values to these propositions, since any body would be in motion or at rest

[28] *Ibid.*, pp. 487–488. I quote Kant at such length in order to include his examples. Very possibly he took them over from Huygens' work on impact forces.

with respect to the co-ordinates of this space. But, at least on Kantian grounds, absolute space is not a possible object of experience. The presupposition fails, and with it the question whether or not bodies in rectilinear "motion" are *really* in motion. Thus Kant says in Proposition 1 of the "Metaphysical Foundations of Phenomenology," the last chapter of the *Metaphysical Foundations of Natural Science* (a chapter that in many respects overlaps the discussion in the Phoronomy) that the rectilinear *motion* of an object is a merely possible (in contrast to a "really possible") predicate.[29]

In his discussion of Phoronomy, Kant sometimes suggests that the relativity of rectilinear motion is of a piece with the "impossibility" of absolute space. We have ourselves been somewhat uncautious in this respect. But in the Phenomenology, he presents the situation in a clearer light. According to Newton, absolute motion entails absolute space. At the same time, Newton accepts the Galilean "principle of relativity" of rectilinear motion. The obvious consequence is that he distinguishes between *absolute* and *relative* motion. In his view, rectilinear motion is indeed relative, while accelerated motion is absolute: "It is indeed a matter of great difficulty to discover and effectually to distinguish the true motions of particular bodies from the apparent. . . . Yet the thing is not altogether desperate; for we have some arguments to guide us, partly

[29] The consideration that rectilinear motion is "relative" undoubtedly provided Kant with a motive for insisting on the (transcendental) subjectivity of space. But it is difficult to see how the subjectivity of space *follows from* the relativity of motion. Vuillemin, *Physique*, p. 59, says that "c'est la relativité du mouvement qui rend transcendentalement necessaire la subjectivité de l'espace." But, first, the relativity of rectilinear movement is not, as we shall see in a moment, *incompatible* with the concept of absolute space, and, second, I have suggested that space has a twofold subjectivity: it is "subjective" from the point of view of the Aesthetic insofar as it is *a priori*, i.e., the subjectivity of space explains how it is that all objects of (outer) experience must be spatial; it is "subjective" from the point of view of the Analytic insofar as the "determination" of space depends on the activity of the understanding. Admittedly, the Transcendental Exposition of the Concept of Space in the Aesthetic blurs the line between these points of view, for it uses the premise that mathematics has a precise "fit" to the world in arguing that space must be subjective, i.e., *a priori*.

from the apparent motions, which are the difference of true motions; partly from the forces which are the causes and effects of the true motions. . . ."[30]

Since, according to Newton's first law of motion, a body not subject to forces persists in its state of (uniform, rectilinear) motion, accelerated motions are caused by forces. It would seem to follow that if we could identify the forces we could identify the accelerated motions, hence identify absolute motion and make a case for absolute space. Kant's response is to distinguish between *absolute* and *true* motion, as Newton apparently did not. There is no reason, Kant suggests, why we could not determine which of two bodies accelerating with respect to each other was *really* in motion, by measuring the forces involved. But to say that one of the bodies is really in motion is *not* to say that it is in absolute motion. "In other words, the question is about the true motion, in contradistinction to illusion, but it is not about the motion as absolute, in contrast to relative."[31] Rectilinear motion is apparent, accelerated motion is true; but this does not force us to conclude that accelerated motion is absolute motion. Thus Kant denies one of Newton's premises, that a body is in true motion if and only if it is in absolute motion. For Kant, both rectilinear and accelerated motion are relative insofar as all motion involves change of spatial relations, insofar as all motion is determined with respect to the "sensible measures" we possess of it.

This is not quite the end of the matter. Newton thought he could demonstrate the existence of motion, even though there "was nothing external or sensible" with which the bodies in motion could be compared, and hence demonstrate the existence of absolute space. ". . . if two globes, kept at a given distance one from the other by means of a cord that connects them, were revolved about their common center of gravity, we might, from the tension of the cord, discover the endeavor of the globes to recede from the axis of their motion,

[30] *Principia,* from the Cajori edition, p. 12.
[31] *MFNS,* p. 561.

and from thence . . . compute the quantity of their circular motions."[32] Kant discusses this case in the last chapter of the *Metaphysical Foundations of Natural Science*, although he uses a different illustration, that of the earth rotating on its axis in infinite empty space. We could, he says, determine that the earth was really in motion by observing various force-effects (e.g., by observing that a stone dropped into a hole reaching to the center of the earth did not fall vertically), but this would not imply the existence of absolute motion. For this motion, he continues, "even though it is no change of relation to empirical space, is nevertheless no absolute motion but a continuous change of the relations of matters to one another. . . ."[33] I think what Kant has in mind (i.e., it is implied by the way in which he discusses the issue) in this passage is that the detection of the various force-effects is in terms of the relations between bodies (e.g., between the earth and the stone); i.e., only absolute force-effects would entail absolute motion, and hence absolute space, but absolute force-effects are undetectable. On this reading of the argument, the epistemological considerations that he directs against the "possibility" of absolute space are brought to bear on absolute motion and eventually on absolute force-effects. The underlying premise seems to be that insofar as anything—space, motion, forces—is perceptible it must involve the application of "sensible measures."[34]

[32] *Principia*, p. 12. There is an excellent discussion of Newton's argument in Bas van Fraassen, *An Introduction to the Philosophy of Time and Space* (New York: Random House, 1970), pp. 108–116. See also R. S. Westfall, "Newton and Absolute Space," *Archives internationales d'histoire des sciences*, 67 (1964), pp. 121–132, and H. M. Lacey, "The Scientific Intelligibility of Absolute Space: A Study of Newtonian Argument," *British Journal for the Philosophy of Science*, 21 (1970), pp. 317–342.

[33] *MFNS,* p. 561.

[34] More support for this claim, and hence for the interpretation offered in this chapter, will be provided in chapters 6 and 7. I will have more to say about the existence of forces, and their determination, in the next chapter.

In fact, Kant seems to have been somewhat unsure about what to say in detail about Newton's case. He observes, for example, in the Metaphysical Foundations of Phenomenology (pp. 557–558) that:

"on the present subject, one can refer to the latter part of Newton's

objectivity concepts

I want now to rejoin the main argument of this chapter and make some remarks about the location of the Axioms of Intuition in the Analytic. A closer look than we have so far taken at the argument of the second edition of the *Critique* is needed.

I begin with one of the notes which Kant appends to the Aesthetic in the second edition. He there asserts that ". . . everything in our knowledge which belongs to intuition . . . contains nothing but mere relations; namely, of locations in an intuition (extension), of change of location (motion), and of laws according to which this change is determined (moving forces). What it is that is present in this or that location, or what it is that is operative (*wirke*) in the things themselves apart from change of location, is not given through intuition. Now a thing in itself cannot be known through mere relations; and we may therefore conclude that since outer sense gives us nothing but mere relations, this sense can contain in its representation only the relation of an object to the subject, and not the inner properties of the object in itself" (B66).

The usual comment on this passage is that Kant is underlining his insistence on the transcendental ideality of space and time and repeating his view, against Leibniz, that the "inner natures" of objects are unknowable. This is all right as far as it goes, but it does not go far enough. In addition to claiming that insofar as empirically real, space and time are properties

---

scholium to the definitions with which he begins his *Mathematical Principles of Natural Philosophy*. From this it will become clear that the circular motion of two bodies around a common center (and hence also the rotation of the earth on its axis) even in empty space, and hence without any possible comparison through experience with external space, can nevertheless be cognized by means of experience, and that therefore a motion, which is a change of external relations in space, can be empirically given, although this space itself is not empirically given and is no object of experience. This paradox deserves to be solved."

But it is not clear that the passage on page 561 is intended as a resolution of this "paradox" (it certainly was not a paradox from Newton's point of view), nor is it at all clear how this passage could be used to resolve it.

of appearances and not things in themselves, and *a fortiori* that things in themselves do not have, despite Newton and naive realism,[35] a metric, Kant also suggests that our knowledge of spatial and temporal properties depends on relations that are themselves physical. In particular, measurement of space and time presupposes the existence of physical processes, for example motion (which in turn presupposes the existence of moving forces as a way of making the distinction between true and apparent motions), which have certain properties.[36] In the case of time measurement, one of these properties would be that the processes are repeatable and invariable.

The point about *knowledge* extends in another direction. For Kant, knowledge is equated with objective experience and this in turn with experience of objects. The question he poses in the Analytic is "what conditions must be satisfied if our experience is to be construed as experience of objects?" That is, what conditions must be satisfied for knowledge to be possible? These conditions involve the possession and application of concepts that, given their purpose, we may well call "objectivity concepts." They serve to mark out a distinction between subjective and objective experience, between experiences of which an individual is the subject and a world of which at least some of these experiences are experiences. The Axioms figure in the Analytic, then, as objectivity concepts.

This may be brought out in two ways. On one, note that whereas the Aesthetic marks out a contrast between the transcendental and empirical reality of space and time, the Analytic allows for a contrast between objective and temporal experi-

[35] Despite Newton's protests to the contrary, it seems that his view is essentially that of naive realism, which takes quantities to be intrinsic and nonrelational. Kant's position is that quantities are relational; one cannot separate "real quantities" from their "sensible measures and relations" or "abstract from our senses and consider things in themselves."

[36] See, for example, the *Critique*, B156: "for all inner perceptions we must derive the determination of lengths of time . . . from the changes which are exhibited to us in outer things"; and the *MFNS*, p. 556: "a relation, and hence also a change of this relation, i.e., motion, can be an object of experience only insofar as both of motion's correlates are objects of experience; but pure space, which is also called absolute space in contradistinction to relative (empirical) space, is not object of experience and is nothing at all."

ence, *within* experience. But one of the ways in which objective and subjective experience is to be distinguished is in terms of the measurability of space and time. Kant demands of the object of experience that it have a precise location in space and time. There are, by way of contrast, no precise spatial relations between objects considered simply as objects of the visual fields of different people. There is no requirement that subjective visual fields stand in any determinate spatial relations to one another. But the public space (and time) that embraces the objects we perceive is of a determinate character. Among other things, it is measurable. This is presumably one of the reasons why those who have distinguished between "primary" and "secondary" qualities of objects in terms of the measurability of the former have insisted on their "objectivity" as against the "subjectivity" of the secondary qualities.

The other way in which we might show how the Axioms function as objectivity concepts depends upon the argument Kant adds in the second edition of the *Critique*: "Appearances, in their formal aspect, contain an intuition in space and time, which conditions them, one and all, *a priori.* They cannot be apprehended, that is, taken up into empirical consciousness, save through that synthesis of the manifold whereby the representations of a determinate space or time are generated, that is, through combination of the homogeneous manifold and consciousness of its synthetic unity. Consciousness of the synthetic unity of the manifold (and) homogeneous in intuition in general, in so far as the representation of an object first becomes possible by means of it, is, however, the concept of a magnitude (*quantum*)" (B202-203). This proof derives more from the argument of the Transcendental Deduction than from the Aesthetic. We learn in the Deduction that all "taking up into empirical consciousness" requires a synthesis of the manifold. But, in part, this synthesis of the manifold involves the successive representation of its parts. So, since intuitions cannot be apprehended except through a succession of their parts, they are extensive magnitudes; this is what it means to be an extensive magnitude.

From the standpoint of the "Subjective Deduction" in the

first edition of the *Critique*, all apprehension involves synthesis. The unity of consciousness requires that in an important respect we "construct" the objects of our experience. From this standpoint, the proof of the Axioms turns on the fact that our "construction" involves successive synthesis, and hence that the objects we "construct" are extensive magnitudes.[37] We are reminded also that *quantities* are not "given"; they are "constructed" by some of the same acts of synthesis that "construct" the objects that have them. But the point can be made independent of the talk about "acts of synthesis." It is simply that objects are objects of experience, "really possible" objects, on the condition that we conceptualize them in certain ways. In particular, objects are objects of experience insofar as they are determinate, measurable. This is a fundamental condition of objectivity. But the measurability of objects, and the possibility of taking them up into empirical consciousness, depends ultimately on certain conceptual abilities of ours.

Kant's argument revisited

One way of viewing Kant's enterprise, we have seen, is as an argument from some body of knowledge assumed as given, Euclidean geometry or Newtonian mechanics, for example, to the "presuppositions" or "*a priori* conditions of the possibility" of that knowledge. These "presuppositions" are statements that must be true if the statements that presuppose them are to have a (knowable) truth value: *A* presupposes *B* just in case whenever *A* is true *B* is true and whenever *A* is

[37] It should be pointed out that it is misleading, in any case, to say that the argument of the Axioms goes like this: our generation of magnitudes, e.g., of a line, is successive, therefore all magnitudes are extensive (or, perhaps more accurately, that since our awareness of objects is successive, all appearances are extensive). For this is to put the point backwards. As I have tried to emphasize, Kant's argument is that we know from the Transcendental Deduction that the unity of consciousness requires the application of objectivity concepts, in particular as a way of distinguishing between objective and subjective temporal experience; one of these concepts is that of extensive magnitude (for determinate temporal magnitudes require that temporal magnitudes be extensive).

false *B* is true. In the case at hand, measurement presupposes the Axioms of Intuition, assignment of magnitudes to motions presupposes that they can be added. Unless we assume satisfaction of the appropriate topological and metrical conditions, measurements cannot be carried out. In other words, statements assigning numbers to objects, for instance, that projectile *a* has terminal velocity Ø, have no truth value. But the "possibility" of a mathematical physics, of which Newtonian physics is the paradigm for Kant, depends on such measurements.

Another way of viewing Kant's enterprise is as an argument from certain alleged facts about the unity of consciousness to the Categories as necessary conditions of that unity. The unity of consciousness—from one point of view, the possibility of distinguishing between my experiences and a world of which they are experiences—entails application of the Categories.[38] Once again in the case at hand, the Axioms are a necessary condition of the unity of consciousness insofar as they make possible the concept of a public, objective world in which individuals may locate themselves precisely.[39]

On my interpretation, then, there are two different sorts of arguments at stake. One is an argument to the effect that natural science (by which Kant means in every instance the application of mathematics to a grouping of natural phenomena) presupposes certain *a priori* principles. The other is an argument to the effect that the unity of consciousness entails the application of the Categories. I shall call the former the "presupposition" argument and the latter the "necessary condition" argument. The fact that there are two different sorts of argument is obscured by Kant's remark, already quoted, that the *Prolegomena* were composed according to the "analytic" or "regressive" method, the *Critique of Pure Reason* according

[38] Properly speaking, the "schematized" Categories or "Principles."

[39] After-images, for example, are both spatial and temporal, but they do not occupy positions in a determinate spatial-temporal framework, which is to say that they are neither public nor objective. But to say that they do not occupy positions in a determinate space-time framework comes, as far as the present point is concerned, to saying that they are not measurable.

to the "synthetic" or "progressive" method. This suggests that the difference between the two is one of "direction" only, and that the arguments are otherwise perfectly symmetrical. But, as I indicated at the outset, generations of commentators following up this suggestion have come to grief trying to make the arguments that Kant actually gives in the *Prolegomena* and *Critique* compatible with it. A closer look at the *Prolegomena* and the *Critique* reveals that they are not perfectly symmetrical treatments of the same arguments. For one thing, whereas the two arguments have separate sets of premises, the existence of natural science and the unity of consciousness, respectively, they have the same conclusion, that certain *a priori* principles, the Categories, find application to our experience. For another thing, if we "ascend" to the Categories from one "direction" and "descend" to them from another, then the Categories cannot be unequivocally "necessary conditions" for both science and the unity of consciousness, as some commentators have believed.

Having noted this crucial asymmetry, let me indicate two ways in which the "presupposition" argument and the "necessary condition" argument are connected. The first has to do with the status of the Categories. The "presupposition" argument establishes their *a priori* aspect; the statements of a natural science have no truth value except on the assumption that the Categories are true, that is, find application to our experience. I have tried to illustrate this point here with respect to the Axioms of Intuition and more specifically in the discussion of the Metaphysical Foundations of Phoronomy. But the "presupposition" argument does not guarantee the *uniqueness* of the Categories. Given that there is such a thing as a natural science, that we can determine which mathematically expressed propositions about defined groupings of natural phenomena are true and false, we can isolate its "categorical" presuppositions. But in doing so we have given no reason why we should have just these presuppositions, why only certain sorts of statements descriptive of natural phenomena should have truth values. For many people, such a question, especially in light of the relativistic and quantum mechanical developments, would be unimportant or unanswerable or nonsensical. But not for Kant. He thought there was a way to prove not only that our experience had a definite general character given the fact that it had a definite special (that is, "Newtonian"—with the qualifications to be discussed in the next chapter) character, but that it *had* to have this character. In other words, he thought one could definitively fix the limits of "real possibility," say in advance of experience what any world that we are capable of experiencing must, at least as regards its general features, be like. This is a central purpose of the "necessary condition" argument. The Categories are not only presupposed by science, they are entailed by the unity of consciousness. This brute fact of consciousness, the "deepest" fact to which a philosopher could possibly appeal, ensures the necessity of the Categories. In some very strong sense, short of their being logical truths, there can be no alternative to them.[40]

Kant has a third argument,[41] for the *completeness* of his list of Categories, which should also be mentioned. We might know that the Categories as synthetic *a priori* principles made experience possible, in the sense that they were presupposed by natural science, and that the Categories were necessary, in the sense that they were entailed by the unity of consciousness, without knowing whether a particular list of Categories was complete. Kant tries to guarantee the completeness of his list by appealing to the Aristotelian table of judgment forms from which he claims the Categories may be derived. But this move is notoriously weak. Both the "completeness" of the Aristotelian table and the "derivation" in question are suspect.

The second way in which the "presupposition" argument and the "necessary condition" argument are associated has to

[40] I suggested a supplementary argument in an earlier chapter. On that argument, the necessity of a Category, isolated, for example, by the "presupposition" argument, is shown by ruling out its alternatives, the alternatives specified in every case by the law of excluded middle. E.g., either space is relational or it is not, either every event has a cause or it does not, etc.

[41] In many respects, the one he was fondest of.

do with the notion of objectivity. For Kant, objective experience is always experience of objects. The point is that the synthetic *a priori* principles presupposed by science as the conditions of its possibility are at the same time constitutive of a world of objects independent of any particular experience, or group of experiences, of them. In the case of the Axioms of Intuition, this comes to saying that the objects are precisely located vis-à-vis one another in a spatial-temporal framework. Further, it is the idea of a world of objects that allows us to distinguish our private, objectless experiences from the common experience of an independent world in which we all participate. I will try to develop this theme in more detail in the next three chapters.

Chapter 5: Kant and Newton

NEWTON'S name is as inextricably connected with Kant's theory of science as is Euclid's. Usually we are told that Kant began with a belief in the validity of Newtonian physics.[1] But "Hume's sceptical attack on the validity of causal inference—and thereby on the possibility of all empirical knowledge"[2]—made a philosophical defense of Newton's theory necessary. What had to be done was to show that, in spite of Hume, causal inference is valid. Indeed, Kant did just this, and a great deal more besides. He showed that Newtonian physics can be derived from certain unquestionable premises having to do with the fact of consciousness.[3] In this way, Kant secured Newton's theory against all possible objections, by giving it a firm metaphysical foundation. For ". . . it is a *consequence* of Kant's metaphysics of experience that Newton's theory is valid."[4]

The argument is often turned around. Since Newton's theory has been shown, first by the theory of relativity and then by quantum mechanics, to be invalid, and since it is a consequence of Kant's metaphysics, there must be something wrong with the metaphysics. Friendly critics want to say that Kant's position must be placed in historical perspective.[5] Kant

[1] Kemp Smith: "Newton, he believes, has determined in a quite final manner the principles, methods and limits of scientific investigation." *A Commentary on Kant's 'Critique of Pure Reason,'* p. v.

[2] R. P. Wolff, *Kant's Theory of Mental Activity*, p. 25.

[3] Lewis White Beck: "The *Critique of Pure Reason* . . . does not assume science and mathematics but rather establishes, by a general epistemological inquiry, principles from which they may be derived." Introduction to Beck's translation of the *Prolegomena* (New York: Liberal Arts Press, 1950), p. xvii, note.

[4] W. Stegmüller, "Towards a Rational Reconstruction of Kant's Metaphysics of Experience (I)," *Ratio* (1967), p. 15.

[5] R. G. Collingwood: "The truth is that the Transcendental Analytic is an historical study of the absolute presuppositions generally recognized in Kant's own time and as a matter of fact for some time afterwards. . . . Some of these go back to Galileo. Some of them are today fallen into desuetude. If

succeeded in isolating the underlying assumptions of Newton's theory, no mean achievement. It is sometimes added, indeed, that this is all a philosopher can and should do: elaborate the presuppositions of particular theories. More hostile critics contend that the whole enterprise is mistaken if we think of these assumptions or presuppositions, even with respect to individual theories, as synthetic *a priori* propositions. There just are not any synthetic *a priori* propositions, as the overthrow of Newtonian physics should lead us to realize. If any propositions are presupposed, in the sense that they are not subject to empirical verification or falsification, then these propositions are analytic, for example, the propositions of logic and mathematics.

There are a great many issues mixed together here. It will be one of my main aims to sort them out. The discussion is organized around four topics: the nature of Hume's challenge, Kant's response to the challenge, the bearing of the challenge on physics, and the relations between Kant's theory and Newton's. Only some very general points about this last topic will be made here. In succeeding chapters I will try to examine the relation between physics and particular Categories in more detail. For the present, my central purpose is to show how and why the standard sketch of Newton's connection with Kant must be revised.

Hume's challenge

What did Kant take Hume's challenge to be? In the first place, he took it to be a challenge not to physics but to metaphysics. As he says, for example, in the *Prolegomena* (Academy edition, pp. 256–257): "Since Locke's and Leibniz's Essays, or rather since the beginning of metaphysics so far as the history of it reaches, no event has occurred which could have been more decisive in respect to the fate of this science than the attack which David Hume made on it." In refuting Hume,

---

the unity of the whole constellation is insisted upon, there is nothing for it but to say that it forms a set of absolute presuppositions not actually made as a whole until Kant's lifetime, which lasted for about a century after he formulated it. . . ." *An Essay on Metaphysics*, p. 245.

Kant hopes to save metaphysics. For Hume's objection can be generalized. ". . . the concept of the connection of cause and effect is by no means the only one by which connections between things are thought *a priori* by the understanding; indeed . . . metaphysics consists of nothing else whatever." If Hume's argument is sound, there simply is no such thing as metaphysics.

One of the difficulties for the claim that Kant's project is to provide metaphysical foundations for Newtonian science, and thereby prove its validity, is that, to my knowledge, Kant himself never characterizes it in that way.[6] Far from guaranteeing physics from skeptical attack, the task is to say how metaphysics can become, like physics, a science. This is Kant's project, as the full title of the *Prolegomena* makes clear. The emphasis is particularly strong in the two prefaces to the *Critique of Pure Reason*. In the first, he says: "We often hear complaints of shallowness of thought in our age and of the consequent decline of sound science. But I do not see that the sciences which rest upon a secure foundation, such as mathematics, physics, etc., in the least deserve this reproach. On the contrary, they merit their old reputation for solidity, and, in the case of physics, even surpass it," and adds immediately that a critique of pure reason is needed to decide "as to the possibility or impossibility of metaphysics in general, and determine its sources, its extent, and its limits—all in accordance

[6] Heinrich Scholz, in the lectures referred to in the Preface, formulates (p. 11) what he calls Kant's third "physical-theoretical principle" as follows: "Eine Physik von der Struktur der klassischen Physik ist die einige Physik, die für die Konstituierung der wirklichen Welt in Betracht kommt. Folglich kann die wirkliche Welt auf keine Art erweitert werden zu einer Welt, die beherrscht ist von irgen welchen Gesetzen, die in der klassischen Physik nicht vorkommen können." But Scholz is careful to add in a note that his principle was not "explicitly formulated by *Kant*."

I was very much heartened to read a few pages later (p. 14) that Scholz seems to favor, although he does not develop, the same thesis I will advance in this chapter: "Die Gesamtheit der Urteile im Sinn der Kantischen Urteilslehre kann aufgefasst werden als der Übergang von einem postivistisch interpretierten Wirklichkeitsbegriff und einer positivistisch interpretierten Physik zu einer realistisch interpretierten Wirklichkeitsbegriff und einer realistisch interpretierten Physik."

with principles" (Axii). In the second preface it is urged that "Metaphysics . . . though it is older than all other sciences, and would survive (i.e., as a 'natural disposition,' not as a science) even if all the rest were swallowed up in the abyss of an all-destroying barbarism, it has not yet had the good fortune to enter upon the secure path of a science" (Bxiv).

Now how did Kant think Hume's argument challenged metaphysics? Roughly as follows. Hume showed that it is never self-contradictory to affirm the antecedent and deny the consequent of particular causal judgments.[7] In other words, using Kant's terms not Hume's, causal judgments are synthetic. Take any traditional metaphysical principle. In the same way it can be shown that the principle is synthetic. But, the second premise of the argument, if a principle is synthetic it must, once again using Kant's terms not Hume's, be *a posteriori*. The principle of causality, Hume insists, is derived from experience. Since, third premise, metaphysical principles are, "by definition," *a priori*, it follows that there are no metaphysical principles.

Kant's response

Kant's response to the argument is, of course, to deny the second premise.[8] It does not follow from the fact that a proposition is synthetic that it is *a posteriori*. There is another possibility: a proposition might be both synthetic and *a priori*.

Kant sometimes suggests that Hume simply overlooked this possibility and that it is enough to mention it to invalidate

[7] This is the argument of the *Enquiry Concerning Human Understanding* and not the *Treatise of Human Nature*. Although there is some evidence to indicate that Kant was familiar with the discussion of causality in the *Treatise* through his reading of Beattie, in his reply Kant seems to have the *Enquiry* version of Hume's argument in mind. For he accuses Hume of mistakenly inferring the contingent character of the principle of causality from the contingent character of particular causal judgments. See the *Critique*, A766/B794: "Hume was therefore in error in inferring from the contingency of our determination *in accordance with the law* the contingency of the *law* itself."

[8] As he makes explicit at B127: From the premises he used, Hume "argued quite consistently. It is impossible, he declared, with these concepts and the principles to which they give rise, to pass beyond the limits of experience."

Hume's argument. At other times, he suggests that the *possibility* of synthetic *a priori* propositions is entailed by the fact that there *are* synthetic *a priori* propositions, as at B128, for example: "Now this empirical derivation (of Hume's) cannot be reconciled with the scientific *a priori* knowledge which we do actually possess, namely, *pure mathematics* and *general science of nature*; and this fact therefore suffices to disprove such derivation." But Kant was also aware that this move is not satisfactory. Kant's problem was to account for the fact that *a priori* propositions, produced by reason, were synthetic in that they had application to the objects of our experience, i.e., were "really possible." It does not help very much to be told that there are such propositions. A solution to Hume's problem involves an explanation. We must know how the "real possibility" of synthetic *a priori* propositions can be guaranteed.

When Kant asks "how are synthetic *a priori* propositions possible?" he is to be understood as asking how, or under what conditions, synthetic *a priori* propositions can be said to have application. This is equivalent, for him, to asking under what conditions they are meaningful, under what conditions they have a (knowable) truth value. There is no problem, presumably, in connection with synthetic *a priori* propositions. Their application is guaranteed by the fact of their derivation from experience. This is no doubt one of the reasons why Hume thought that if a proposition was synthetic it had to be *a posteriori* as well; no other conclusion would allow the fit between our concepts and the world to be explained. Neither is there a serious problem, from one point of view, in connection with the synthetic *a priori* propositions of mathematics.[9] For in mathematics, as we have seen, the objects to which our propositions apply have been constructed by us in such a way that a fit is guaranteed. The problem is to say how objects that are in some sense "given"—that is, which are not a product of the mathematical (or any other) imagination—can be de-

[9] Note that Kant takes all non-mathematical synthetic *a priori* propositions as "metaphysical."

scribed or determined *a priori*, as metaphysics pretends.[10] If, like God, we were capable of constructing objects for all our concepts, there would be no problem here either. But human intuition is not productive in the same way. There is, Kant rightly insists, a passive element in all perception.

the "objectivity" of Newtonian physics

The foregoing remarks are not intended to deepen anyone's understanding of Hume or Kant. They are only reminders of positions with which I expect most readers of this book are already familiar. But these remarks should put us in a position to understand the bearing of Hume's challenge on Newton's theory. On my reading, Hume's challenge, if unmet, precludes a *realist* interpretation of Newton's theory. What Kant is concerned to show is not that Newton's theory can be "established," to use a frequent expression, on the basis of certain fundamental synthetic *a priori* principles, with or without the addition of experimental observations, and hence guaranteed against skeptical attack, but rather that it is "objective." To say that it is "objective" is to say that it applies to objects that are independent of us in the sense (a) that these objects occupy determinate spatial-temporal locations, (b) that they are capable of causally interacting with one another (and in the special case of causing us to have sensory reactions to

[10] It is clear from the letter to Marcus Herz of February 21, 1772, that this was the problem that in large part motivated the *Critique of Pure Reason*. In his *Inaugural Dissertation*, Kant had argued that metaphysical propositions find application to things in themselves, independent of our "sensible modes" of experiencing them. On turning to a closer examination of the issue, he saw that no way of justifying the application of such propositions to things in themselves could be found. In fact, the very concept of a thing in itself precludes the possibility of saying anything (knowledgeably) about it. The question then became: can the application of metaphysical propositions to appearances, that is, objects insofar as they are sensibly experienced by us, be justified? The *Critique* was intended to provide an affirmative answer. The solution to the corresponding problem in the case of mathematics had been worked out as early as the Berlin prize essay of 1764. In a way, Kant's final solution to the "problem of metaphysics" involves generalizing the results of this essay to the "pure" part of physics.

them), (c) that they continue to exist when unperceived by anyone, and (d) that they are not necessarily perceivable with the unaided senses alone, as in the case of the "magnetic matter pervading all bodies."[11] This is going to require some spelling out.[12]

To begin with, why should Hume's challenge preclude a realist interpretation of physics? Suppose that what is "given" are sense-impressions. How, on the basis of these sense-impressions, can we claim that, for example, objects continue to exist when unperceived? A possible answer is that we make our claim on the basis of causal laws. We can "reach" any object, so to speak, given present sense-impressions and laws that allow us to "project" these impressions. On the other hand, if the appropriate causal laws cannot be justified rationally, as Hume seemed to have shown, then rational projection is out of the question and our claim about the continued existence of unperceived objects, to use the same example, unjustified.

Having reached this point in the argument, Hume suggests two different moves. The more sophisticated move is to say that the gaps in our sense-experience created, for example, by the assumption that objects continue to exist unperceived (or that there are objects unperceivable by the unaided senses), are filled in by the useful fictions of the imagination. We "feign" the existence of objects to fill in the gaps. This comes to taking physical objects as "theoretical constructs" that

[11] As we shall see in chapters 6, 7, and 8, there is a close connection between an object's having a spatial-temporal location and its having a specifiable range of causal properties.

[12] Note that the differences between my reading and that sketched at the outset of this chapter are blurred by using the word "valid." Thus, when someone says that Kant wanted to prove that Newtonian physics is "valid," he may mean (a) that Newtonian physics necessarily applies to the natural world, (b) that Newtonian physics follows as the conclusion of some argument, (c) that Newtonian physics is true, (d) that Newtonian physics applies to the world (i.e., that the subject terms of its propositions denote really possible objects). On my reading, Kant wants to prove validity only in this fourth, weakest, sense; his main concern is to guarantee, not the adequacy of the physics, but the existence of the world which it is taken to describe.

serve to explain but are not reducible to patterns of sense-experiences. Our belief in the existence of a world of independent objects cannot be rationally justified, but it can in some sense be "explained."

The less sophisticated, and much more widely imitated, move is to say that, since the gaps cannot be filled in, we should reconstruct our concept of an independently existing object in such a way, in terms of actual and hypothetical patterns of sense-impressions, that the gaps are not created in the first place. On this "phenomenalist" version of Hume, objects are "reduced" to sets of sense-impressions, physical laws—Newton's among them—are reinterpreted so as to be mere instruments of prediction of the course of our actual sense experience. Crudely, to say that whenever *A* then *B* is, on this view, to say no more than that whenever I have a sense-impression of type *A* it will be followed by a sense-impression of type *B*.[13]

For both alternatives, there is an important sense in which Newton's physics is left untouched. All of Newton's theorems hold; they are still "valid." They have just been reconstrued. As Berkeley might have put it, the scientific results have not been doubted. Only the philosophical interpretation of them differs. Berkeley was as thoroughgoing a Newtonian as the next man.[14] So was Hume, as the subtitle

[13] As, for example, Berkeley in the *Principles of Human Knowledge*, #50: "To explain the phenomena is all one has to show, why upon such and such occasions we are affected with such and such ideas."

[14] See, *inter alia*, his letter to Samuel Johnson of November 25, 1729, in *The Works of George Berkeley*, edited by A. A. Luce and T. E. Jessop (London: Thomas Nelson and Sons, Ltd., 1949), II, p. 279: "The true use and end of Natural Philosophy is to explain the phenomena of nature; which is done by discovering the laws of nature, and reducing particular appearances to them. This is Sir Isaac Newton's method; and such method or design is not in the least inconsistent with the principles I lay down. This mechanical philosophy does not assign or suppose any one natural efficient cause in the strict and proper sense; nor is it, as to its use, concerned about matter, nor is matter connected therewith; nor doth it infer the being of matter." Berkeley did think there were metaphysical elements, for instance the notion of absolute space, in Newton's work that could be excised without damage to the latter. To excise the "metaphysical" elements, particularly the concept of matter, is

of the *Treatise* and the final paragraph of the *Enquiry* make abundantly clear.[15] We can say, following philosophical tradition, that independently existing physical objects (as I have characterized them) have or are "matter." Berkeley and Hume want to purge physics of matter. When Kant couples skepticism and empiricism, as he often does, the reference is to skepticism about the existence of external objects, and in particular to the existence of imperceptible particles, skepticism about the existence of "matter," not about the possibility of developing a "science of nature."

Both sorts of Humean purgatives are rejected by Kant.[16] He wants a fully realist, or *material*, interpretation of Newto-

---

to provide what might be called a *minimal interpretation*, that is, in terms of what is immediately experienced, of Newton's theory.

[15] It might be pointed out that Hume's critique of "natural philosophy" is directed, as is Berkeley's, against a particular philosophical interpretation of it. See, for example, the *Treatise*, Book IV, Part IV, Section IV, "Of the Modern Philosophy," where Hume argues that the scientific distinction between primary and secondary qualities cannot be taken to correspond to a real distinction in objects, and the *Enquiry*, Section IV, Part II, where doubts raised about induction concern only the conjunction between sensible qualities of objects and what Hume calls their "secret powers." To give a realist interpretation of physics, as I understand it, is in part to accept the primary/secondary quality distinction, which Kant does (see the *Critique*, A45/B62), and the inference from an object's sensible qualities to its presumed micro-structure, which Kant also does (see A226/B273). In fact, there is a close connection between the view that objects have a micro-structure in terms of which their behavior is to be explained and the view that only the primary qualities are real (for Kant, read "empirically real") properties of objects. Presumably both Berkeley and Hume thought that these realist elements in Newton's own position, as expressed, for example, in his "Rules of Reasoning in Philosophy" (particularly the third) in *Principia*, could be pared away without harm to his "experimental laws."

[16] Kant does not seem to have appreciated Hume's "sophisticated" move, that physical objects are theoretical constructs introduced to explain although not reducible to patterns of sense-impressions. In certain ways, Hume's use of the "feigning" activities of the imagination resembles Kant's doctrine of *a priori* synthesis; both allow us to go "beyond" our fragmentary, momentary, disconnected experience to a world of persisting, interacting, unified physical objects. Despite this fact, I think it is a mistake not to see important differences between them. Some of these will be indicated as we go along, particularly in the final chapter.

nian physics and he thinks that one can be guaranteed by the argument of the Transcendental Deduction. The argument is complex and has received a great deal of commentary. But, put in its simplest terms, I think it comes down to this: unless we assume that those *a priori* principles that make a world of independent objects possible have application, from which follows the possibility of a realist interpretation of physics, then we will not have an adequate concept of the self. In a way, Kant takes the contrapositive of Hume's argument in the Appendix to the *Treatise.* An adequate concept of the self, as subject of experience, requires a concept of an object as having definite spatial-temporal location. But for us to be able to make out what it is for an object to have a definite spatial-temporal location requires that certain *a priori* principles—causality, substance, and community, for example—have application to our experience. Briefly: the conditions that have to be satisfied if physics, and mathematics, are to have application, that is, objective reality, are the conditions that must be satisfied if there is to be a "world," that is, a set of objects having definite spatial-temporal locations. And these conditions must be satisfied if we are to have an adequate concept of the self, that is, a concept of the self as subject of experience. What follows from what Kant calls "the unity of consciousness" is not Newtonian physics, or the "validity" of Newtonian physics, but rather the possibility of providing a realist or material interpretation of Newtonian physics. Kant's great insight in the Transcendental Deduction is that the unity of consciousness amounts to the possibility of being able to distinguish between what is objective and what is subjective in our experience and that this possibility in turn requires more than a *minimal* interpretation of Newton's (or any comparable) theory. It is not just that the *a priori* principles that make a more robust interpretation possible do have application. They must.

Before we go any further in this direction, it might be a good idea to round up more direct support for my claim that Kant is interested not in defending Newton but in providing a realist interpretation of his results.

We can begin by returning to a piece of text I have already discussed, the first of the three notes to #13 of the *Prolegomena*. Kant there says that

"Pure mathematics, and in particular pure geometry, can only have objective reality under the condition that it bears merely on the objects of the senses. . . . From this it follows that the propositions of geometry are not determinations of a mere creature of our poetic fantasy which could not be reliably referred to real objects. . . . The space of the geometer would be held to be mere fiction and would be allowed no objective validity; because it cannot be seen how things must agree necessarily with the picture that we make of them by ourselves and in advance."

The same dual problem arises in the case of physics, unsurprisingly, since for Kant physics is no more than applied mathematics: to say what the conditions are that must be satisfied if the propositions of physics are to have objective reality and to show that those conditions are satisfied.

Kant continues in the same passage: "It will always remain a remarkable phenomenon in the history of philosophy that there was a time when even mathematicians who were also philosophers began to doubt, not indeed the correctness of their geometrical propositions in so far as they merely concern space, but the objective validity and application to nature of this concept itself and of all geometrical determinations of it." Newton himself might very well be intended. In *Principia*, there are elements of two very different views. Alongside the realist view, expressed, for example, in the pronouncement that "gravity really exists . . . (it is an) active principle," there is a *formalist* view to the effect that *Principia* is no more than an exercise in advanced mathematics. The formalist elements invariably come to the fore when the reality of forces is at stake (exception made for pronouncements like the one just quoted concerning gravity). Newton says, for instance, that in *Principia* he intends

". . . only to give a mathematical notion of those forces, without considering their physical causes and seats; (and that quite generally, he is) considering those forces not physically, but mathematically; (that he does not want to) define the kind, or the manner of any action, the causes or physical reasons of those forces; (does not want to attribute) in a true and physical sense (forces) to certain centres (which are only mathematical points); when at any time . . . (he happens) to speak of centres as attracting, or as endowed with attractive power."[17]

On the basis of this sort of passage, at least one of Newton's early reviewers was led to say that *Principia* was "mathematics," not "physics." But, for Kant, it is physics; hence the need to demonstrate the reality of forces. For Berkeley and for Hume, on the other hand, force-talk is translated into motion-talk, dynamics reduces to kinematics.[18] "Forces" drop out of the picture of the world, to be replaced by generalizations about motion that are themselves analyzed in terms of regular sequences of sense-impressions. As a result, there is no problem about the *application* of Newton's physics. The "fit" of mathematics to the world is assured ultimately by the "reduction" of both objects and the propositions about them to sense-impressions. In a very different direction, Descartes similarly guarantees "fit" (apart from the unsatisfactory appeal to God's benevolence) by eliminating forces; objects are "reduced" to their geometrical properties, matter is identified with extension, and force-talk is explained away in terms of the perceived deviations in the movement of objects.

[17] *Principia*, Cajori edition, pp. 5-6. For a discussion of the formalist-realist ambiguities in Newton's position, see Gerd Buchdahl's very interesting article, "Gravity and Intelligibility: Newton to Kant," in R. E. Butts and J. W. Davis, eds., *The Methodological Heritage of Newton* (Toronto: University of Toronto Press, 1970). Richard Westfall, in the most extended discussion of Newton's views on the concept of force with which I am acquainted, *Force in Newton's Physics* (New York: American Elsevier, 1971), argues for a realist interpretation (see, e.g., pp. 509-510).

[18] See Berkeley's *Siris*, in *The Works of George Berkeley*, ed. A. C. Fraser (Oxford: Clarendon Press, 1901), III, pp. 231—"The laws of attraction and repulsion are to be regarded as laws of motion . . ."—through 235—"We are not therefore seriously to suppose, with certain mechanic philosophers, that the minute particles of bodies have real forces or powers, by which they act on each other, to produce the various phenomena in nature."

But, again, it is just this sort of "reduction" of matter that Kant resists. Hence, this time against Descartes, it once more becomes a question of establishing the reality of forces.

The problem of the "objective reality" of physics is pointed up in at least three additional ways. One of these concerns Kant's attack on idealism. What serves to unite most of his philosophical predecessors, both rationalist and empiricist, is their common idealism, which is to say that in one form or another they doubt or deny the existence of independent objects having definite spatial-temporal locations. As we shall have occasion to see shortly, these doubts and denials are intimately bound up with questions having to do with the reality of forces.[19] But to be an "idealist" is, in the case of physics, just to deny the possibility of a realist or material interpretation of it. On the other hand, to make out the necessity of such an interpretation is to "refute" idealism. And it is clear that Kant is bent on refuting idealism, not only from the section entitled "The Refutation of Idealism," where Berkeley and Descartes (and by implication the two traditions of which they are representatives) are grouped together as "idealists," but throughout the *Critique*, as in the preface to the second edition:

"However harmless idealism may be considered in respect of the essential aims of metaphysics (though, in fact, it is not thus harmless), it still remains a scandal to philosophy and to human reason in general that the existence of things outside us (from which we derive the whole material of knowledge, even for inner sense) must be accepted merely on *faith*, and that if anyone thinks good to doubt their existence, we are unable to counter his doubts by any satisfactory proof" (Bxl, note a).

A second way to point up Kant's concern with a realist interpretation of physics is to recall his insistence on his own empirical realism and his corollary claim that only his own

[19] Leibniz's view is complex in this respect and not so easily characterized. There is a sense in which he is an "idealist" and yet takes the reality of forces as fundamental.

position, so-called "transcendental idealism," makes empirical realism possible.[20]

A third way to point up Kant's realist concern is to look at his characterization of his project as a transcendental inquiry. He tells us at A11/B25 of the *Critique* that all knowledge is entitled transcendental "which is occupied not so much with objects as with the mode of our knowledge of objects insofar as this mode of knowledge is to be possible *a priori*." I take this to imply that a transcendental inquiry is not concerned with the correctness of, for instance, Newtonian physics, but rather with the philosophical interpretation to be placed upon it. To put it in a slightly different way, Kant's use of "transcendental" here is designed to make a sharp distinction. perhaps for the first time in the history of thought, between scientific and philosophical questions. The distinction was not, according to Kant, respected by his predecessors. It was the mistake on which traditional metaphysics rested. As a result, his predecessors tended to embark on the project of finding philosophical foundations for natural science.[21] But once we realize that as philosophers we are interested not in objects but in concepts and the conditions of their application we shall realize that it is not a philosophical task to "establish" a particular physical theory.

Kant's use of "transcendental" has another aspect. We can think of the language of a particular theory as an object language. Then a "transcendental" inquiry, philosophy as rightly understood, will be a meta-theoretical investigation relative to that object language. A transcendental inquiry will establish, for example, what the singular terms of the theory

[20] See the Paralogisms of Pure Reason in the first edition of the *Critique of Pure Reason*, at A371 for example: "The transcendental idealist is, therefore, an empirical realist, and allows to matter, as appearance, a reality which does not permit of being inferred, but is immediately perceived" and footnote 13 to chapter 1.

[21] Even Hume: "In pretending therefore to explain the principles of human nature, we in effect propose a compleat system of the sciences, built on a foundation almost entirely new, and the only one upon which they can stand with any security." *A Treatise of Human Nature*, edited by Selby-Bigge (Oxford: Clarendon Press, 1888), p. xx.

in question are, under what conditions they may be said to refer, etc.—in short, will provide a commentary on the semantic development of the theory. In arguing for a realist interpretation of physics, Kant is indicating at the same time what he takes to be the philosopher's task and maintaining as well that only a particular semantic commentary is adequate.

matter, nature, and Newton

In the next two chapters, I shall have more to say in detail about the bearing of particular Categories on the possibility of providing Newtonian physics with a realist interpretation. But first I want to try to clarify my general thesis, in part by taking a closer look at Kant's program in the *Metaphysical Foundations of Natural Science.*

Notice, to begin with, that nothing I have said so far in attempting to reconstruct Kant's position commits us to the necessity of Newton's theory. It does not follow from the claim that we can and ultimately must give a realist interpretation of that theory that alternative theories are thereby ruled out. We are not even committed to the truth of Newton's theory. For nothing I have said so far precludes the possibility that under a realist or "material" interpretation all of Newton's propositions might turn out to be false.[22]

Kant is perfectly explicit that the propositions of physics are contingent. "We cannot," he says, "without destroying the unity of our system, anticipate general natural science, which is based on certain primary experiences" (A171/B213). Only such experiences, in contrast to mere ratiocination, can determine which of its propositions are true and which false. From this it follows that the propositions of physics are not synthetic *a priori.* For if they were *a priori*, then of course experience would not be necessary to determine their truth or falsity.

Unfortunately, Kant is not always consistent. At B17 of the

[22] By an "interpretation" here I mean providing referents for the singular terms of the theory. I do not mean providing a model for the theory; in that case, of course, all the propositions would come up true.

Critique, we are told that "natural science contains *a priori* synthetic judgments as principles." He then cites two such judgments: "that in all changes of the material world the quantity of matter remains unchanged" and "that in all communication of motion, action and reaction must always be equal." The first looks suspiciously like a principle of conservation of matter, the second like Newton's third law of motion. This is not very satisfactory for the following reasons. First, if Newton's third law is synthetic *a priori*, then why not the first and second laws? And if the three laws are synthetic *a priori*, and if we take them as adequate for the axiomatic development of Newton's theory, then in what way is an appeal to experience necessary to determine the truth values of the propositions of physics? Second, Kant says that all non-mathematical synthetic *a priori* propositions are metaphysical. If the propositions of Newton's theory are synthetic *a priori*, then there is, again contrary to Kant's own pronouncement, no distinction to be made between physics and metaphysics. Third, the passage at B17 of the *Critique* does not square with #15 of the *Prolegomena*. At B17, Kant says that the two propositions mentioned above belong to the "pure" part of natural science. But in #15 of the *Prolegomena*, he says that although these same propositions belong to "general natural science" they are not "wholly pure and independent of sources in experience."[23] The propositions of "pure" natural science *there*

[23] In the Introduction to the Critique, at B3, Kant distinguishes between "pure" and "impure" *a priori* propositions. An *a priori* proposition that contains none but *a priori concepts* is pure. The propositions of (pure) mathematics and the first Analogy, that substance is permanent, might be given as examples. An *a priori* proposition that contains *empirical concepts*, such as "All red roses are red," is impure. Thus, the "laws of motion" listed by Kant in the *Metaphysical Foundations of Natural Science* are *impure*, since they contain the concepts of "matter" and "motion" that are empirical (i.e., can only be given in experience).

Even in the Introduction, however, Kant's account does not seem to be quite consistent. At B3 he says that "the proposition, 'every alteration has its cause,' while an *a priori* proposition, is not a pure (*rein*) proposition, because alteration is a concept which can be derived only from experience," whereas at B4 (a scant two paragraphs later) this same proposition is used as an example of a pure (*rein*) *a priori* judgment. The twin contrasts, *a priori/a posteriori*, mentioned are "substance remains and is permanent" and "everything that happens is always previously determined according to constant laws by a cause." These, he says, really are universal laws of nature that subsist wholly *a priori*. Furthermore, these (properly metaphysical-looking) principles are the ones "deduced" as principles of the possibility of experience in the Analytic as the first and second Analogies of Experience.

The criterion of consistency, placed on any rational reconstruction of a philosopher's work in the Preface, would suggest that we follow the *Prolegomena* here rather than the *Critique* and make a distinction between "pure" natural science on the one hand and physics, including Newton's theory, on the other hand. But the criterion of consistency is not always easily invoked. I think the slippage in Kant's position apparent from the passages quoted stems not from carelessness but from a deep tension in the position. We are better advised to try to explain it than to try to explain it away.

Let us, then, take a closer look than we have done so far at the one work of Kant's "mature" period devoted especially to the philosophy of physics, the *Metaphysical Foundations of Natural Science*. Kant's program in the *MFNS*, to put it very briefly, is to "construct" the concept of matter, a concept he takes to be at the center of natural science, and thereby to guarantee its "real possibility." This construction[24] at one and the same time demonstrates the mathematizability of the concept of matter, and thus secures the application of mathematics to nature, and proves its objective reality. To say this is, of course, to recapitulate the argument of the preceding four chapters.

Two closely connected motives seem to have guided this

---

pure/empirical, simply do not allow Kant to draw all the distinctions between types of propositions that he wants to draw, although as I suggest in the text there are underlying philosophical reasons for his tendency to slide at this point.

[24] A spatial representation of matter as the movable. Construction is always with respect to the forms of intuition, largely with respect to space.

program. One concerns Kant's "refutation of idealism" and has already been introduced. Kant's "refutation" turns on the claim that the unity of consciousness requires the existence of objects in some sense external to us. It cannot merely be the case that such objects have spatial location, for otherwise they are not to be distinguished from volumes of empty space. Spatial location does not by itself provide us with a suitable empirical criterion for the existence of objects external to us. In addition to spatial extension, we must also attribute forces to such objects, the possibility of their causally interacting with one another and, of utmost importance, with ourselves. As Kant puts the point at A265/B321 of the *Critique*: "We are acquainted with substance in space only through forces which are active in this and that space, either bringing other objects to it (attraction), or preventing them from penetrating into it (repulsion and impenetrability). We are not acquainted with any other properties constituting the concept of the substance which appears in space and which we call matter."[25] Thus the "construction" of the concept of matter has to do in the first place with a distinction between matter and space, contra Descartes,[26] hence with the necessity of attributing forces to objects. In this respect, the "construction" of the concept of matter has to do in the first place with the completion of Kant's enterprise in the *Critique of Pure Reason.*[27]

The other motive that seems to have guided Kant's program in the *Metaphysical Foundations of Natural Science* has to do more specifically with what might be termed the problem

[25] See also A277/B333, A285/B341, A371, A413/B440, and A618/B646 ("in fact, extension and impenetrability [which between them make up the concept of matter] constitute the supreme empirical principle of the unity of appearances").

[26] And J. H. Lambert, Kant's friend and frequent philosophical correspondent, who vigorously defended the Cartesian identification of matter and space.

[27] In this connection, Kant makes a very interesting remark in the *MFNS*, 478: ". . . it is indeed very remarkable . . . that general metaphysics in all cases where it requires instances (intuitions) in order to provide meaning for its pure concepts of the understanding must always take such instances from the general doctrine of body . . . the understanding is taught only through instances from corporeal nature what the conditions are under which the concepts of the understanding can alone have objective reality, i.e., meaning and truth."

of forces.[28] Again, one aspect of the problem has already been introduced in earlier references to Berkeley and Hume. The "problem of forces" centered in, although it was not limited to, difficulties with the notion of action at a distance. There was no question about the inductive support for the law of universal gravitation, or for the laws of motion; that is, they were predictively confirmed. But there was a philosophical difficulty with the *intelligibility* of action at a distance. Newton's philosophical friends Locke and Clarke maintained that action at a distance is "a contradiction" and "impossible to conceive," and Leibniz hammered away at the theme that it was at best an occult quality. Even Newton himself had tried, without success, to eliminate action at a distance in favor of a mechanical hypothesis. Kant denies that action at a distance is logically impossible,[29] but he is faced with the problem of its "real possibility."[30] This is one of the problems Kant hopes to solve in constructing the concept of matter, as well as the more general problem having to do with the reality of forces (whether propagated at a distance or not).

As I have said, what Kant is trying to do in the *Metaphysical Foundations* is to demonstrate the "real possibility" of the concept of matter. Now to demonstrate the "real possibility" of a concept is not merely to analyze it; it is to give the conditions for its application. Thus a condition for the application of spatial concepts is that they have an empirical representation. In the *MFNS*, Kant first tries, on analogy with the argument of the Axioms of Intuition and the Anticipations of Perception in the *Critique*, to show under what conditions the concept of matter is mathematizable; this involves establishing an appropriate additivity rule (given in the law of composition of velocities) and solving various problems connected with continuity. At the level of the Phoronomy, or kinematics, there are no important difficulties. At this level, matter can be represented as points. There is no distinction to be made between

[28] The following paragraph owes a great deal to Buchdahl's article, "Gravitation and Intelligibility," although he develops the material in a very different direction.

[29] *MFNS*, 513–514.

[30] Compare Berkeley: attraction "is only a mathematical hypothesis, and not a physical quality." *De Motu*, in the Luce and Jessop *Works*, IV, p. 38.

matter and space. The physics of bodies is indistinguishable from geometry. It is perhaps just this fact, I suggested earlier, that assures the application of geometry in the first place.

Kant then tries to prove, more on analogy with the Analogies of Experience,[31] the "objectivity" of the concept of matter. This involves, in part, setting out the conditions that must be satisfied if matter is to be distinguished from empty space. In particular, the "objectivity" of the concept of matter is shown to depend on our attributing two fundamental forces to matter (these are not "included in" its concept but, as conditions of its "possibility," "belong to" it). Questions concerning the construction of the concept of matter then become questions concerning the construction of these two fundamental forces.

Kant's program in the *Metaphysical Foundations of Natural Science*, then, might be summarized in this way. While the physicist alone can determine the specific properties of different sorts of objects, the philosopher can determine what general properties its concept must have if it is to be a "possible" concept. In the case at hand, the philosopher can determine, *a priori*, that each piece of matter must be endowed with attractive and repulsive forces, but only the physicist can determine the particular magnitude of these forces.

If not always clear in detail, the main outlines of the program are comprehensible. But Kant's attempt to carry it out reveals two sorts of difficulties buried in the program. One of these difficulties originates in his claim that the concept of matter is an empirical concept[32] because an element in the concept, motion, cannot be "cognized" *a priori*. Nevertheless, the concept of matter seems to play a rather curious role, somewhere between purely *a priori* and purely empirical con-

[31] The parallel breaks down in certain respects. The Metaphysical Foundations of Dynamics, the second chapter of the *MFNS*, mixes together questions having to do with the application of mathematics—e.g., continuity, on analogy with the Anticipations of Perception—and questions having to do with the "objectivity" of the concept of matter—e.g., the proofs that matter must be endowed with two fundamental forces.

[32] The concept of matter is empirical, but the proof of its "possibility" (requiring application of the transcendental method) must be philosophical.

cepts (Kant says, in fact, that it has certain *a priori* elements *in it*). On the one hand, the concept is empirical insofar as the "possibility" of motion cannot be demonstrated *a priori*; motion can be given only *a posteriori*, in experience. On the other hand, the possibility of experience, and ultimately the unity of consciousness, seem to require something like the concept of matter, first for the construction, i.e., empirical representation and determination, of space and time, then for making out a distinction between objects and space, on which not only the construction just mentioned, but also the "refutation of idealism" would seem to depend. Kant's narrow dichotomy—*a priori* or *a posteriori*—does not allow him to deal adequately with the concept of matter.[33] At the same time, the fact that the concept of matter is called on to play different roles on different occasions accounts for the shifts in Kant's attitude toward the *a priori/a posteriori* status of Newton's theory.

I am not sure how this difficulty is to be resolved without widening Kant's framework so as to include a place for, and an explanation of, concepts like matter that are not either *a priori* or *a posteriori*. I think it would help, however, to distinguish between a "strong" and a "weak" position on the concept of matter. The "strong" position is that matter really is an *a priori* concept. Any "really possible" world must be one in which the concept of matter finds application. The "weak" position is that in *our* world the spatially extended permanent required by the unity of consciousness is *in fact* identical with matter. Matter happens to function as the perceptible representation of spatial and temporal relations, but (perhaps) we can conceive of other sorts of change besides movement in terms of which these relations could be defined. At least, the

[33] Kant sometimes recognizes the peculiarly "intermediate" position of the concept of matter by distinguishing, in #15 of the *Prolegomena*, for example, between pure natural science (the Principles), general natural science (the propositions demonstrated in the *Metaphysical Foundations of Natural Science*), and physics proper; propositions having to do with the possibility of the concept of matter fall under general natural science (*physica generalis*). See Hansgeorg Hoppe, *Kants Theorie der Physik* (Frankfurt am Main: Vittorio Klosterman, 1969), for a discussion of the way in which commentators have traditionally disagreed about the role and status of the concept of matter.

"weak" position is compatible with the reconstruction of Kant's theory of science as I have developed it so far, although I have also tried to indicate how the motives guiding Kant's program in the *Metaphysical Foundations of Modern Science* could have pushed him in the direction of the "strong" position. What seems to me to be crucial is that there are obvious differences between the Categories and Newton's laws in Kant's respective treatment of them (the former required for the unity of consciousness, the latter not) which lumping them together as synthetic *a priori* principles only obscures. In any event, whatever necessity Newton's laws have is not by virtue of the fact that they imply or are implied by the Categories, but because they articulate conditions that, at least in our world, must be satisfied if the concept of matter is to find application.

The other sort of difficulty in Kant's attempt to carry out his program concerns the construction of forces. Since the concept of matter, on Kant's analysis of it, essentially contains attractive and repulsive forces,[34] the construction of the concept eventually involves the construction of these forces. But since according to Kant our knowledge of these forces is inevitably *a posteriori*,[35] they cannot be constructed. Thus, the task Kant sets himself in the *MFNS*, to construct the concept of matter, ends in half-admitted failure. I say "half-admitted" because there is a certain amount of hedging on his part. On the one hand, for reasons I have already indicated, attractive and repulsive forces "make possible the *general* concept of matter."[36] On the other hand, owing to their *a posteriori* aspects, it is not possible to "construct this concept (in detail and thus) represent it as possible in intuition."[37] Even more confusing is Kant's suggestion[38] that although *he* has failed to construct (completely) the fundamental forces, and hence the concept of matter, there is still left open the possibility that they might be constructed by others.

[34] This is not quite accurate. Such forces are not "contained in" the concept of matter, but "belong to" it and "make it possible," for reasons to be more fully explored in the next chapter.

[35] *Metaphysical Foundations of Natural Science*, 524.

[36] *Ibid.* [37] *Ibid.*, p. 525. [38] *Ibid.*, p. 534.

Hans Reichenbach makes the interesting remark that "Kant's concept of *a priori* has two different meanings. First, it means "necessarily true" or "true for all times," and secondly, "constituting the concept of an object."[39] It is possible that part of our present difficulty in reconstructing a consistent Kant is owing to his failure to distinguish sharply enough between these two meanings and his correlative failure to distinguish between "matter" as that which is *there* over and above the forms we impose on it and "matter" as it is conceived in classical physics, i.e., "mass." Kant seems to have thought that just as the Categories (implicitly) define the concept of an object in general, a generalized version of Newtonian mechanics defines the concept of an object in particular, and that the latter *had* to apply (had, in some sense, to be *a priori*) if the former were to have any content. One thing is clear: Kant thought that he could establish that "pure" natural science, as he characterized it, applied necessarily and for all time. And it is at this point that we, having passed through at least two conceptual revolutions in physics since the 18th century, must break with him. What has happened, it seems fair to say, is that the classical spatial-temporal framework, and the concept of matter on whose application that framework somehow depended for Kant, have changed or, perhaps more accurately, have been replaced.[40]

"Pure" natural science, and especially the intermediate

[39] *The Theory of Relativity and A Priori Knowledge*, translated by Maria Reichenbach (Berkeley: University of California Press, 1965), p. 48.

[40] Again, this same point seems to have been made by Reichenbach (who uses the expression "principle of coordination" in the following passage in roughly the way that Kant uses "synthetic *a priori* judgment"): "The principles of coordination represent the rational components of empirical science at a given stage. This is their fundamental significance, and this is the criterion that distinguishes them from every particular law, even the most general one. A particular law represents the application of those conceptual methods laid down in a principle of coordination; the principles of coordination alone define the knowledge of objects in terms of concepts. Every change of the principles of coordination produces a change of the concept of object or event, that is, the object of knowledge. Whereas a change in particular laws produces only a change in the relations between particular things, the progressive generalization of the principles of coordination represents a development of the *concept of object* in physics." *Ibid.*, pp. 87–89.

"general physics," as conceived by Kant, is now seen not so much to be false as to be irrelevant. But as a result the propositions of classical physics no longer have truth values; in some sense they are now "meaningless." Thus, "the (classical) mass of x is Ø" is now not simply false; it is neither true nor false since the classical concept of mass no longer applies to objects as relativistic physics construes them. In the same way, "the precise spatial-temporal location of *a* is *x, y, z, t* and its velocity is *s*" is neither true nor false, for once again the "presuppositions" of classical physics no longer obtain, this time vis-à-vis quantum mechanics.[41] Perhaps it is in coming to realize the inadequacy of Kant's claim that a particular set of presuppositions could be guaranteed once and for all that we realize the truth of another claim, that at any given time science consists of *a priori* as well as empirical elements.

the law of universal gravitation

Towards the end of his discussion in the *Prolegomena* of the second main transcendental question, "how is pure natural science possible?" Kant once again takes up the theme of the "law-giving" power of the understanding, which in turn suggests a "rationalist" interpretation of his position. Since I have been urging a rejection of that interpretation, and of an oft-advanced corollary that for Kant Newton's laws of motion are both synthetic and can be known *a priori*, it would be advisable to take a brief look at this discussion.

In #36 of the *Prolegomena*, Kant asserts (as he does elsewhere, particularly in the first edition of the *Critique*) that *"the understanding does not draw its laws (a priori) from nature, but prescribes them to nature"* and then proceeds (in #38) to clarify this "apparently daring" assertion by way of an example. The example is drawn from geometry. Consider the theorem that "two lines, which cut each other and (a) circle, . . . always divide each other regularly, so that the rectangle erected on the segments of either line is equal to that on the segment of the other." It is Kant's claim that this theorem "can be de-

[41] See the final section of Karel Lambert's paper, "Logical Truth and Metaphysics," in Lambert, ed., *The Logical Way of Doing Things* (New Haven: Yale University Press, 1969).

duced solely from the condition which the understanding placed at the ground of the construction of this figure, namely the equality of the radii." This, of course, is the familiar claim that geometrical theorems follow from postulates that are essentially "constructive" in character. In the example given, the Axioms of Intuition (and the metric concept of equality) are clearly at stake: the theorem is already implicit in the "successive synthesis of the productive imagination in the generation of" a circle. Given the validity of geometry, it is in this way that the understanding determines the general properties of space and in this sense that it prescribes "laws" to nature.

Kant elaborates his point by applying it to a fundamental law of physics, Newton's law of universal gravitation.

"If we go still further, to the fundamental doctrines of physical astronomy, we find a physical law of reciprocal attraction extending over the whole material nature, the rule of which is that it decreases inversely with the square of the distances of each attracting point, just as the spherical surfaces in which the force diffuses itself increase, which seems to lie necessarily in the nature of things themselves and hence is usually propounded as capable of being known *a priori*. Simple as are the sources of this law, in that they rest merely on the relation of spherical surfaces of different radii, its consequence is so excellent in respect of the variety and regularity of its agreement, that it follows not only that all possible orbits of heavenly bodies are conic sections, but that they have such a relation to each other so that no other law of attraction than that of the inverse square of the distances can be conceived."

Now, in the first place, Kant does not want to say that the LUG is a synthetic *a priori* proposition or that it has Categorial status. So much is indicated by Kant's calling it a "physical" proposition and his saying that the test implications so well conform to the facts.[42] Furthermore, if the LUG is not synthetic *a priori*, then it is difficult to see how the laws of mo-

[42] See the *MFNS*, 534: ". . . no law whatever of attractive or repulsive force may be risked on a priori conjectures; but everything, even universal attraction as the cause of gravity, must, together with the laws of such attraction, be concluded from data of experience."

tion could be. But, in the second place, the law does "rest" on universal principles of the determination of space. Kant evidently has Newton's proof of the law in mind. Newton begins his proof by showing that if the path of an object is a conic section and if the force on the object is always directed to one focus, the force must be inversely proportional to the square of the object's distance from the focus. Thus, an object whose orbit describes an elliptical path, as in Kepler's first law of planetary motion, is acted on by a force F that at any given moment is proportional to $1/R^2$, where R is the distance from the focus. The point is that the proportionality of the force to the inverse square of the radius is a property of conic sections, and hence follows as a theorem from the postulates that lay down the conditions for the construction of conic sections. It is in this way that the understanding is prescriptive, with respect to the form of experience. Once space has been "determined" by the understanding, through application of the Principles, then once we have shown that the orbit of an object is a conic section (and whether or not its orbit is a conic section is a matter of empirical fact; it is in no way necessary), the object *must* obey the inverse square law. In this sense, physics presupposes the Principles. For physics is possible only when experience, space and time in particular, has a determinate structure.[43] In this same sense, the law of universal gravitation "stands under" the Principles.

[43] Thus, in commenting on the above quoted passage from the *Prolegomena*, Kant says: "Space is something so homogeneous and in respect of all particular properties so indeterminate that there is certainly no hoard of natural laws to be found in it. On the other hand, that which determines space into the figure of the circle, the cone and the sphere, is the understanding, in that it contains the ground of the unity of the construction of these figures. The mere universal form of intuition that is called space [i.e., what I called "spatiality" in chapter 3] is indeed the substratum of all intuitions which can be designated to particular objects, and admittedly there lies in space the condition of the possibility and variety of intuitions. But the unity of objects is determined solely by the understanding, and according to conditions which lie in the nature of the understanding. Thus in that it subsumes all appearances under its own laws and only by doing so brings into being experience (as to its form) *a priori*, the understanding is the origin of the universal order of nature. . . ."

Chapter 6: the substance of matter

THE ever-present danger in giving a rational reconstruction of a philosophical position is that the search for what makes sense—inevitably taken as a function of current practices and principles—leads one too far away from the original text. I mention it here because the danger seems to me to be especially close when commenting on the Analogies of Experience. The argument in this section of the *Critique of Pure Reason* is compressed; there are many different issues, and the impression of non-sequitur and inconsistency is great. For these same reasons, however, the text of the Analogies provides a kind of two-fold test of any reconstruction, of its accuracy on the one hand and of its ability to clarify on the other.

The framework of interpretation developed so far allows at least a preliminary clarification of some traditional difficulties. Take the first Analogy, for example. Kant calls it the "Principle of Permanence of Substance." His point is not to prove that substance is permanent, this is analytic or, as he says, "tautological." Rather, for motives we are now in a position to understand, what he wants to show is that this principle must find application to our experience. In the first edition of the *Critique*, it is formulated as follows: "All appearances contain the permanent (substance) as the object itself, and the transitory as its mere determination, that is, as a way in which the object exists." And in the second edition: "In all change of appearances substance is permanent; its quantum in nature is neither increased nor diminished."

The difficulty is that these seem more like a formulation of two different concepts of substance than like two formulations of one concept. On the first concept—call it the Aristotelian concept—substance is the substratum of change, that of which properties can be predicated but which cannot be predicated of anything else. On the second or Cartesian concept, substance is that which exists in its own right, depending for its existence on itself alone, uncreated and indestructible. We might rephrase the matter to bring the two concepts into closer connection: on the "Aristotelian" view, substance is the invariant in all change and process; on the "Cartesian" view, substance as the invariant is that which remains eternally the same. But even this close connection is not close enough. The concept of something that has properties, even the substratum of change, is distinct from the concept of something that is conserved over time.

Things become somewhat clearer in light of our isolation of two asymmetrical patterns of argument that Kant uses. One, the so-called "presupposition" argument, seems related to the Cartesian concept of substance, for it might be supposed that conservation is a "presupposition" of classical physics; except on the assumption that such conservation principles hold, there is no way to insure appropriate closure of the physical systems we observe, hence no way to measure changes in them. The other, so-called "necessary condition," argument seems related to the Aristotelian concept of substance, for it might be supposed that a distinction between a substance and its properties (and, it develops in the argument of the Analogies of Experience, as a consequence the possibility of distinguishing between objective and subjective temporal relations) is required by the unity of consciousness. Of course, we still have a long way to go. The connection between each concept and its corresponding pattern of argument has to be made more precise, and the connection between them shown, for Kant obviously thought of them as intimately connected. In a word, both of them have to do with the question of "objectivity." In this chapter, I will say something about the first Analogy, placing the emphasis on the "presupposition" argument. This will involve taking a still closer look at the *Metaphysical Foundations of Natural Science*. In the next chapter, I will say something about the second Analogy, underlining the way it functions as a "necessary condition" of the unity of consciousness. Various connections between the two Analogies will be noted along the way.

substances and space

Kant gives several arguments for the necessary application of the principle of the permanence of substance to the objects of our experience, all of them brief, none of them completely satisfactory. In the main, they revolve around three premises: "All appearances are in time; and in it alone, as substratum (as permanent form of inner intuition), can either coexistence or succession be represented" (B225); "Time cannot by itself be perceived" (B225); "All existence and all change in time have thus to be viewed as simply a mode of the existence of that which remains and persists" (A183/B227).

The first premise is relatively unproblematic. Kant takes it to have been demonstrated in the Aesthetic that neither "coexistence nor succession would ever come within our perception, if the perception of time were not presupposed as underlying them *a priori.*" To say of two events that one took place after the other, or that they took place at the same time, implies established use of the concept of time. It is not "reducible" à la Leibniz or Hume.[1] In this sense, time is an *a priori* concept; all temporal discriminations require it as a background against which they can be made. But there is no background, on pain of infinite regress, against which time can itself be dated or discriminated.

The second premise—"that time cannot by itself be perceived"—though it can be misleading, is also relatively unproblematic. What Kant means is that the empirical determination of temporal relations must be in terms of perceptible changes, that is, in terms of events or, equivalently, some change in our perceptual experience. To borrow a phrase from the *Inaugural Dissertation* (#14), the quantity of time cannot be measured "except in the concrete." This seems true, although it is perhaps not self-evident (as Kant seems to think). But to estimate quantity of time in the concrete is to relate it to events or happenings that are perceptible. Thus,

[1] Recall the parallel discussion of the irreducibility of the concept of space in chapter 3.

for instance, two stretches of time are identical just in case they are overlapped by all and only the same events. In other words, making temporal discriminations requires that changes take place. A world in which there are no changes is a world in which there is no empirical discrimination of time.

The third premise is the heart of the argument. It says that events, so far taken in a wide sense, must be construed as changes in objects, rather than, for example, mere successions of perceptions. If they are not, then an objective determination of time relations will not be possible. Unless we refer changes in our perceptual experience to changes in the objects of experience, there will be nothing to guarantee a distinction between objective and subjective temporal relations on which the unity of consciousness depends.

Putting it this way brings the argument of the first Analogy directly into the long line of argument beginning with the Transcendental Deduction. I will have more to say about it from this point of view in the next chapter. But even if one were not inclined to buy the whole Deduction-Analogies argument, there are still good reasons for thinking that events in every case are changes in objects. Strawson emphasizes two of these in his book *Individuals*: that the possibilities for identifying and locating events without reference to objects are limited,[2] and that events are conceptually dependent on objects in the sense that we could not have, say, the idea of a birth or a death without the idea of an animal that is born or dies. More generally, we could not conceive of an event, a change, that was not a change in something that itself did not change, a substance. Change implies, at least insofar as it implies a distinction between change and that in or to which change takes place, permanence.[3]

[2] Events do not provide "a single, comprehensive and continuously usable framework" of reference of the kind provided by objects. *Individuals* (London: Methuen & Co., Ltd., 1959), p. 53.

[3] I do not mean to suggest that event-talk can be "reduced" to object-talk, only that there is a kind of conceptual dependency. Indeed, there are reasons for thinking that objects are, in turn, conceptually dependent on events. See Donald Davidson, "The Individuation of Events," in Nicholas Rescher *et al.*, eds., *Essays in Honor of Carl G. Hempel* (Dordrecht: D. Reidel Publishing Company, 1969).

There is still an important gap in the argument as I have sketched it. Kant acknowledges this gap in the last paragraph of the section on the first Analogy. After restating the intended conclusion—"Permanence is thus a necessary condition under which alone appearances are determinable as things or objects in a possible experience" (i.e., as being related to one another in time)—he adds: "We shall have occasion in what follows to make such observations as may seem necessary in regard to the empirical criterion of this necessary permanence—the criterion, consequently, of the substantiality of appearances" (A189/B232). Two considerations motivate the addition. On the one hand, there is Kant's general dictum that if a concept, of any kind, is to have sense (*Sinn*) and reference (*Bedeutung*) there must be an empirical criterion for its application. On the other hand, he insists in the argument of the first Analogy that since time is not perceivable, it must have a perceptible representation. What is the perceptible representation? Changes in substances. But then what is it for a substance to be perceptible?

Let me put this another way. I have already suggested that the identification of events is derivative: we locate the object the change in which constitutes the event, and thereby locate the event. If this location procedure is going to be very helpful, it cannot be in terms of the temporal locations of objects. For it is *their* temporal location that events make possible in the first place. It must be their spatial location. The location of events via objects depends on these objects occupying spatial positions. In Strawson's words, "we must see objects as belonging to, and events as occurring in, an identical, enduring spatial framework."[4] Kant makes much the same point in the Postulates of Empirical Thought, at B291: "When . . . we take the pure concepts of *relation*, we find, firstly, that in order to obtain something *permanent* in intuition corresponding to the concept of *substance*, and so to demonstrate the objective reality of this concept, we require an intuition in space (of matter). For space alone is determined as permanent, while time, and therefore everything that is in inner sense, is

[4] *The Bounds of Sense*, p. 132.

in constant flux." In a marginal note to the first Analogy in his own copy of the *Critique*, he develops another aspect of this point: "here the proof must be so developed as to apply only to substances as phenomena of outer sense, and must therefore be drawn from space, which with its determinations exists at all times. In space all change is motion."[5] Space is generally the permanent "background" against which objects, and eventually events, can be located. But, more specifically, the existence of motion, in the first place change of relations in space, requires it. And motion is, for Kant, the empirical representation of time.[6] Once again, therefore, space ("the enduring spatial framework") is required by the possibility of objective time-relations.

But now the difficulty is to say in what sense space is perceptible. Presumably, if time is not perceptible, then neither is space, for the regress I suggested earlier in the case of time applies equally to the case of space. How then can it function as the perceptible representation of time or that in terms of which motion can be discriminated? Well, space is not perceptible, but spatial objects are perceptible, and these are the "substances" on which the possibility of objective time-relations depends.

But we have only postponed, not resolved, the difficulty. For if we say that substances are substances insofar as they are spatial, then we have no way of distinguishing between objects and volumes of empty space. Both have just the same set of spatial properties.[7] Descartes seems to have recognized this

[5] Quoted by Kemp Smith, *Commentary*, p. 361.

[6] Throughout his discussion, Kant apparently has the solar measurement of time in mind. Which is not to say that he thinks of the sun as any more than relatively permanent, that is, motionless.

[7] This point, suggested by Berkeley, is worked out in detail by Hume in Book I, Part IV, section IV of the *Treatise*, "Of the Modern Philosophy." Roughly put, Hume's argument is that those who distinguish between primary and secondary qualities and who go on to say that the concept of a physical object "contains" only concepts of primary qualities, have not given us an adequate concept of a physical object. For an object must be conceived of as having at least one secondary, or non-relational, property if it is to be distinguished conceptually (and this is our criterion of adequacy, according to Hume) from volumes of empty space which have only spatial, or relational,

as a consequence of his position and, with certain qualifications, to have accepted it.[8] If to say that something is a substance is to say that it is extended, then space must be a substance. There is no conceptual distinction to be made between them. But this position is not really possible for Kant. He has already suggested that we can think of space as empty of substances, an impossibility if space is itself a substance. Not only this, but in the Aesthetic he criticizes the Newtonians for thinking that space was a substance, a thing that was not really a thing; according to Kant, this is a metaphysical absurdity. How, then, can we distinguish substances and space; what non-spatial criterion for substances can be given?

impenetrability and the two "grand hypotheses"

Kant recognized the problem. Midway through the second Analogy he says: "But I must not leave unconsidered the empirical criterion of a substance, in so far as substance appears to manifest itself not through permanence of appearance, but more adequately and easily through action. Wherever there is action—and therefore activity and force—there is also substance, and it is in substance alone that the seat of this fruitful source of appearances must be sought." A substance, then, is that which not only occupies a spatial location, but which is capable of acting. Whatever is extended and can exert force is a substance.[9]

In saying that substances are capable of exerting a force,

properties. See David Armstrong, *Perception and the Physical World* (London: Routledge & Kegan Paul, 1961), chapter 15.

[8] See the *Principles of Philosophy*, ii. Some important qualifications are elaborated by Buchdahl, *Metaphysics and the Philosophy of Science*, pp. 92ff. There are at least two physical objections to the (unqualified) Cartesian *plenum*: how would one account for motion and for the fact that bodies vary widely in density?

[9] Recall that the apparent circle here is broken by distinguishing between the idea of space (spatiality) and space as given (physical space). It follows from the argument of the Aesthetic that we cannot think of objects but as spatial. But the determination of physical space requires a perceptible representation. I take this case to illustrate Kant's general dictum that one must distinguish between the explication of a concept and the explication of the application of the concept to experience. See the *MFNS*, 540.

Kant wants to say in the first place that substances "fill" and not merely "occupy" space. But since to "fill" a space means to resist everything that strives by its motion to press into a certain space,"[10] this comes to saying that substances are impenetrable. The connection between impenetrability and the concept of substance is made explicit in a number of places, for example in the second note to the Refutation of Idealism. It is worth quoting at length:

"Not only are we unable to perceive any determination of time save through change in outer relations (motion) relatively to the permanent in space (for instance, the motion of the sun relatively to objects on earth), we have nothing permanent on which, as intuition, we can base the concept of substance, save only *matter*; and even this permanence is not obtained through outer experience, but is presupposed *a priori* as a necessary condition of determination of time. . . . The consciousness of myself in the representation 'I' is not an intuition, but a merely *intellectual* representation of the spontaneity of a thinking subject. This 'I' has not, therefore, the least predicate of intuition, which, as permanent, might serve as correlate for the determination of time in inner sense—in the manner in which, for instance, *impenetrability* serves in our *empirical* intuition of matter" (B278).

The connection between substance and impenetrability is even more explicit at A284/B340, where Kant equates "an abiding appearance in time" (the perceptible representation of time) with "impenetrable extension."

At this point the discussion in the second chapter of the *Metaphysical Foundations of Natural Science*, the "Metaphysical Foundations of Dynamics," is of relevance. This chapter is intended to parallel the Anticipations of Perception in the *Critique*, although it mixes together "mathematical" questions having to do with continuity (infinite divisibility) and "physical" questions having to do with the reality of matter. It is the latter sort of question that concerns us here. Kant begins where he left off in the *Critique of Pure Reason*, with the

[10] *Ibid.*, p. 496.

claim that impenetrability is the empirical criterion of substance.[11] Just as he first asked himself under what conditions the concept of substance could have application, he now asks himself under what conditions the concept of matter, impenetrable extension (as thus far characterized), could have application. This is the kind of *a priori* activity in which the philosopher, not the physicist, can engage. It has to do with determining the "real possibility" of concepts.

Kant's answer turns on a contrast between two "grand hypotheses" concerning the ultimate nature of matter: the mathematical-mechanical and the metaphysical-dynamical. As he sees it, the principal difference between them is that the mathematical-mechanical assumes, while the metaphysical-dynamical rejects, the absolute impenetrability of matter. To say that matter is absolutely impenetrable is, in turn, to say both that impenetrability is an ultimate property, not further explainable, and that matter is incapable of being further compressed. In rejecting the absolute impenetrability of matter, then, the metaphysical-dynamical hypothesis holds that impenetrability is to be explained in terms of more basic repulsive forces and that there is no point past which an object cannot be compressed.

The picture associated with the mathematical-mechanical hypothesis is familiar. Gross physical objects are composed of hard particles and empty space, and the elasticity, density, and cohesion of these objects is explained in a derivative way as resulting from impact forces and the varying mixture of particles and empty space in an object; e.g., one object is denser than another on this picture if it contains more particles (of matter) per equal unit volume. On the mathematical-dynamical picture, to the contrary, physical objects are centers of fields of force.

[11] As Kant puts it in the *MFNS* (510), repulsive forces first give us a determinate concept of an object in space (an empirically determinable concept). "Thus it is clear that the first application of our concepts of quantity to matter whereby there first becomes possible for us the transformation of our external perceptions into the experiential concept of matter as object in general is founded only on matter's property of filling space."

Kant's argument for the metaphysical-dynamical hypothesis has two sides. The "negative" side is directed against the mechanical hypothesis.[12] It is to the effect that absolute impenetrability, although not self-contradictory, is an "empty" concept, one that *could not* be given in experience. And the same is true of the attendant concept of "empty space."[13] These claims, of course, are corollaries of the general Kantian doctrine that "everything in our knowledge which belongs to intuition . . . contains nothing but mere relations."[14] The implication is that a concept of matter involving the concept of absolute impenetrability would be similarly "empty," and that a physics that employed it would be, although mathematically adequate, without "objective reality." In other words, the "objectivity" of physics precludes the absolute impenetrability of the ultimate material constituents of physical objects.

The positive argument for the dynamical hypothesis is much more difficult to comprehend.[15] In a series of propositions (the details of which are unimportant for my purposes) Kant argues that the "possibility" of matter requires attributing fundamental attractive and repulsive forces to matter. Only on the relative view does the concept of impenetrability have application. In particular, he argues that attraction is necessary, for otherwise any impenetrable body would disperse to infinity under action of its own repulsive forces, that

[12] The structure of the argument is reminiscent of the Aesthetic, with Kant appearing to claim that only two views concerning the ultimate nature of matter are possible, that one is mathematically but not metaphysically adequate, that the other is metaphysically but not mathematically adequate.

[13] The expression "empty space" is ambiguous. One sense is metaphysically legitimate, the other illegitimate. As Kant sets out his view in a note to A431/B459: "*empty space*, so far as it is *limited by appearances*, that is, empty space *within the world*, is at least not contradictory of transcendental principles and may therefore, so far as they are concerned, be admitted." In its legitimate sense, "empty space" is compatible with the existence of an all-pervasive ether. See the *MFNS*, 533ff.

[14] *Critique of Pure Reason*, B66.

[15] At times, in fact, Kant suggests that only a "negative" argument can be given. See the *MFNS*, 524.

the concept of attraction is not self-contradictory,[16] that attempts to reduce attraction to impact forces are circular, and that repulsive and attractive forces adequately explain the phenomena of elasticity, density, and cohesion. What complicates Kant's argument is his claim that the dynamical view remains no more than a "hypothesis" and, in particular, that the "possibility" of forces and hence of matter cannot be demonstrated. On the one hand, it appears as though repulsive and attractive forces are essential to the matter concept. They are not "included in it" but "belong to it" as conditions of its application. On the other hand, it appears that on the dynamical hypothesis the concept of matter cannot be "constructed" nor the "objectivity" of physics be demonstrated.

I think that the apparent conflict in Kant's position can be explained, if not dissolved, in terms of our interpretative framework. The dynamical hypothesis has this advantage over its rival: the concepts of relative impenetrability, of attractive and repulsive forces, are not "impossible" in the same way that the concept of absolute impenetrability is. There are empirical conditions for the application of the former, but not of the latter, concepts. But at the same time, their "real possibility" cannot be demonstrated *a priori*.[17]

[16] Against the views of Leibniz.

[17] In the *Critique of Pure Reason*, Kant says that the concept of attraction is empty. "Thus it is not permissible to invent any new original powers, as, for instance, an understanding capable of intuiting its objects without the aid of the senses; or a force of attraction without any contact; or a new kind of substance existing in space and yet not impenetrable" (A770/B798). Proposition 7 of the Metaphysical Foundations of Dynamics, on the other hand, reads: "The attraction essential to all matter is an immediate action through empty space of one matter upon another." I take it that the apparent contrast is to be explained not in terms of an implausible sudden shift in view, but rather in terms of a reinterpretation of the concept of attraction such that attraction is not propagated across (transcendentally illegitimate) "empty space." Kant certainly takes it as an advantage of his view that it does not necessitate the postulation (in the bad sense) of "empty space" (*MFNS*, 525). In an important footnote to the *MFNS*, 474, Kant remarks that "Newton's system of universal gravitation is well established, even though it carries with it the difficulty that one cannot explain how attraction at a distance is possible." I understand this to mean, first, that the law of universal gravitation is empiri-

"Real possibility" can be demonstrated *a priori* in two different ways. First, one can exhibit *a priori* an intuition corresponding to the concept. This is the way of mathematical construction. Second, one can argue on the basis of transcendental considerations that application of the concept is required by the possibility of experience. This is the way of Kant's metaphysical method. Now there is a sense in which the concepts of the fundamental forces are "*a priori*"; they cannot be "reduced" to anything more basic.[18] This is in part what is meant by calling them "fundamental." On the other hand, an intuition corresponding to these concepts cannot be exhibited *a priori*. Kant claims that such forces cannot be "constructed."[19] Nor can one argue on the basis of transcendental considerations that the application of force concepts is required by the possibility of experience. Although (I understand Kant as saying) they are required as conditions of the application of the concept of impenetrability, impenetrability is no more than what in fact serves as our criterion of substance. It is an empirical, not an *a priori*, concept.

This reconstruction of Kant's discussion in the Metaphysical Foundations of Dynamics has two main consequences. First, all that can be done from a "dynamical" point of view in securing the "objective reality" of physics is to eliminate obstacles to a realist interpretation.[20] If one assumes absolute

---

cally well confirmed, and to that extent there are empirical grounds for the application of the concept of attraction, and, second, that the "real possibility" of the concept cannot be demonstrated. Perhaps it was dissatisfaction with his treatment of attraction in the *MFNS*, where the existence of an ether is merely conjectured, that led him in the *Opus Postumum* to try to *prove* the existence of an ether and thus provide a mechanical account.

[18] E.g., the attempt to "reduce" attraction to impact forces. See the *MFNS*, 524.

[19] See the "Notes" and "Observations" to Proposition 8 of the Metaphysical Foundations of Dynamics. Kant claims, for reasons that are not entirely clear, that the fundamental forces cannot be "presented in intuition," i.e., represented spatially. The implication is that there are limits to the extent to which such forces can be mathematized.

[20] Kant says in the *MFNS*, 534: "This is all that metaphysics can ever accomplish for the construction of the concept of matter, and hence on behalf of the application of mathematics to natural science respecting the properties

impenetrability, then, since absolute impenetrability is not a possible object of experience, one can never go beyond the mathematical formalism of physics to its physical interpretation. From this point of view, absolute impenetrability provides Berkeley and Hume with arguments for a positivist reading of Newton.[21]

Second, if the propositions of the Metaphysical Foundations of Dynamics are synthetic *a priori*, they are so only in a special sense. They are synthetic insofar as they are not explications of concepts. They are *a priori* insofar as they are explications of the conditions of applications of concepts. But they are not synthetic *a priori* in the sense of the first chapter: they are not true in or of all "really possible" worlds. They do not have the status of the Categories of Pure Reason. Their synthetic *a priori* status has to be "relativized," so to speak, to worlds in which the concept of matter *in fact* has application.[22]

---

by which matter fills a space in determinate measure—namely, to regard these properties as dynamical and not as unconditioned original positions, such, for instance, as a merely mathematical treatment would postulate." On my reading, a merely mathematical treatment leaves it unclear *how* mathematics can be applied to natural science, i.e., how properties of objects not of our own making are nevertheless mathematizable.

[21] Kant suggests, I think, that Newton gets himself into difficulties just because he tries to combine absolute impenetrability and attraction into a single view. Absolute impenetrability implies the existence of (absolutely) "empty space" (at least by way of the corpuscular explanation of density), and "empty space" raises problems for attraction. If we attribute this suggestion to Kant, then we are going to have to read "empty space" as something other than "void" or "vacuum" in passages such as the following: ". . . action at a distance, which is also possible without the mediation of matter lying in between, is called immediate action at a distance, or the action of matters on one another through empty space" (*MFNS*, 511-512).

[22] Given the fact that the propositions in question are conditions of applicability for the term "matter," we could say that they are true in or of every really possible world in which matter exists and in this sense "necessary." Along the lines indicated by Saul Kripke in "Identity and Necessity," in Milton Munitz, ed., *Identity and Individuation* (New York: New York University Press, 1971), these propositions would be both necessary and *a posteriori* (since "matter" is an empirical concept). I am indebted to a referee for this suggestion.

In this same connection, Kant makes a distinction in the Introduction to the

This point can be amplified. Kant's claim that attraction is an essential property of matter has often been construed as part of his general justification of the Newtonian position, and Kant has often been criticized for making what is so clearly an element of the physics of his time into an *a priori* matter.[23] But whether or not Kant is to be criticized for not drawing an accurate line between the *a priori* and the empirical, the situation is more complex. Kant never argues that the dynamical theory of matter is more than a "hypothesis." The attribution of repulsive and attractive forces serves to explain *how* the concept of impenetrability finds application, and in this way our talk about such forces is grounded. But, even granting the validity of this part of Kant's argument, there is nothing "necessary" about the concept of impenetrability, and consequently nothing "necessary" about attractive and repulsive forces. The most that could be claimed is that in a world like ours, in which a particular concept of matter found application, attractive and repulsive forces would have to be assumed as fundamental.

matter and mass

The Metaphysical Foundations of Dynamics spells out one aspect of Kant's concept of substance. The next chapter of the

---

Critique of Judgment between transcendental and metaphysical principles: "A transcendental principle is one by means of which is represented, *a priori*, the universal condition under which alone things can be in general objects of our cognition. On the other hand, a principle is called metaphysical if it represents the *a priori* condition under which alone objects, whose concept must be empirically given, can be further determined *a priori*." From the translation by J. H. Bernard (New York: Hafner Publishing Co., 1968). This is, of course, the sense in which "metaphysical" in the title of the *MFNS* is intended.

[23] See Buchdahl, "Gravity and Intelligibility," p. 101: ". . . Kant's procedure is essentially a conservative one of adapting a given and fairly narrow scientific conception. . . . Thus, the impression is conveyed that the general philosophy provides a *backing* for the theoretical concept, when in fact such a backing is not really forthcoming." Also Mary Hesse, *Forces and Fields*, p. 177: Kant "appears to regard the existence of attractive forces at a distance as an *a priori* truth of the metaphysics of matter. He has, however, clearly imported into the 'metaphysics' empirical considerations other than the notion of matter as impenetrable."

MFNS, the Metaphysical Foundations of Mechanics, spells out another, one more closely related to the "Cartesian" concept of substance I mentioned at the outset. The "Dynamics" is concerned mainly with the concept of impenetrability, the "Mechanics" with the concept of "quantity of matter" or *mass*.

In the Introduction to the *Critique*, one of the examples Kant gives of an *a priori* synthetic principle of natural science is this: ". . . in all changes of the material world the quantity of matter remains the same." The same principle turns up in the *Metaphysical Foundations of Natural Science* as the "first law of mechanics": "in all changes of corporeal nature the quantity of matter on the whole remains unchanged, neither increased nor diminished." The principle of the permanence of substance, on the other hand, runs (in its second edition formulation) as follows: "In all changes of appearance substance is permanent; its quantum in nature is neither increased nor diminished." What is the relation between these two principles?

It is tempting to think that the former is derived from the latter with the help of some empirical concepts. Kant's procedure in the Metaphysical Foundations of Mechanics suggests this. If we take this suggestion seriously, we downgrade the principle of conservation of matter from its synthetic *a priori* status, perhaps, but we preserve the traditional picture of Kant's enterprise: the Categories are presuppositions of science in the sense that they constitute the most general assumptions or first principles that furnish part of the premises for the deduction of the actual content of Newtonian physics. It is in this sense that we are to understand Kant's remarks that the propositions of physics "stand under" or are only "special determinations of" the Categories. In fact, it looks as though the "first law of mechanics" follows in one step from the "principle of permanence of substance," by instantiating on the variable "substance." Apparently Kant has found in the principle of permanence of substance an assumption of physics that is at the same time a necessary condition for consciousness.

As I have filled it in, two things in this traditional picture are left unclear. What sort of argument does Kant think can be given for substance on its "Cartesian" interpretation and why does he claim that conservation of matter is required for the possibility of physics? I sketched an argument for the Aristotelian concept of substance in the first section of this chapter: the concept of an event, and thus the possibility of objective time-relations, requires that we distinguish between changes and that in or to which such changes take place. But this argument does not entail, at least not in any obvious way, that the Cartesian concept of substance must find application to our experience. Nor does the fact, if it is a fact, that the Cartesian concept of substance implies conservation of mass shed any light on Kant's claim that conservation is a synthetic *a priori* principle *because* the "possibility" of physics requires it.

Kant gives an argument for the Cartesian concept in the *Critique of Pure Reason*, but it is difficult to understand: "If some of these substances could come into being, and others cease to be, the one condition of the empirical unity of time would be removed. The appearances would then relate to two different times, and existence would flow in two parallel streams—which is absurd" (A188/B231). Two questions that arise at once are: what is the connection between conservation of substance and its coming into being and ceasing to be, and what is the connection between this latter and the empirical unity of time?

An answer to the first question is provided in Kant's proof of the "first law of mechanics" in the *Metaphysical Foundations of Natural Science:*

"What essentially characterizes the substance, which is possible only in space, and under spatial conditions, and therefore, only as the subject of the *external* sense, is that its *quantum* cannot be increased or diminished without substance coming into existence or being annihilated; for the quantity of an object, which is possible only in space, must consist of parts which are external to each other and these, therefore, if they are real must necessarily be substances."[24]

[24] *MFNS*, 542.

To say that the quantum of a substance has increased or diminished is to say that substances have come into being or passed away, for since substances are necessarily spatial[25] their parts must be spatial as well, hence the quantum of a substance is an *additive* function of the quanta of its parts. But why cannot substance come into being and pass away? Because the conditions set out in the second Analogy would be violated.[26] The coming into being of a substance would initiate a causal chain; i.e., it would be an uncaused event. But this is impossible, since according to the second Analogy every event has a cause. The passing away of a substance would not necessarily be an uncaused event, but it would be an event that brought a particular causal chain to a close. But, again, this is impossible, for every event is connected necessarily to a following as well as a preceding one.[27]

Put in this way, the argument indicates a rather loose connection between the Cartesian concept of substance and the requirements of the unity of consciousness, a connection that depends importantly on a particular reading of the second Analogy. But it does not indicate how the Cartesian concept of substance, still less the "first law of mechanics," functions as a "presupposition" of physics.

I think a clue to Kant's view of the connection between conservation principles and physics is to be found in a remark on his "second law of mechanics," the so-called law of inertia, which states that every change of matter has an external cause. Kant remarks that: "The possibility of a natural science proper rests entirely upon the law of inertia (along with the law of the permanence of substance). The opposite of this,

[25] Argued, for example, in the Refutation of Idealism.

[26] I will comment on the second Analogy in detail in the next chapter.

[27] On this account, what taking contemporary cosmoslogical hypotheses (such as that of Hoyle) that postulate the coming into being if not also the passing away of substance seriously comes to is that we must modify either the additive conception of a substance's quantum or Kant's views of causality. In fact, Kant's argument is not strong enough to show that neither of these can be modified. To the extent that the argument is sound, however, it shows that the adoption of the "continuous creation" hypothesis would lead to fundamental conceptual revisions.

and therefore the death of all natural philosophy, would be hylozoism."[28] The reference to "hylozoism" suggests the following interpretation. In his early work *Thoughts on the True Estimation of Living Forces*, Kant distinguishes between "living" or internal forces and "dead" or external forces.[29] This same contrast is stated as one between determinate or mathematical forces and indeterminate or non-mathematical forces. At this early stage, and in attempting to resolve the controversy between Leibnizians and Cartesians concerning the true measure of *vis viva*, Kant held that both sorts of force should be attributed to objects. But from a Critical point of view, this is no longer possible. For objects are knowable just to the extent that their properties are determinate. In particular, the possibility of a natural science proper, as a system of mathematically articulated laws, requires that its objects be mathematizable. This amounts here to the condition that only "dead" or mathematical forces—i.e., external or impact forces—be considered. This is just the condition stated by the second law of mechanics. In this way the "second law" is a presupposition of a science proper, i.e., of a mathematical physics.

In *Thoughts on the True Estimation of Living Forces*, Kant mentions several different motivations for linking the mathematizability with the "deadness" of forces, including a claim that "dead" but not "living" forces can be deduced from the geometric properties of objects. In the *Metaphysical Foundations of Natural Science*, the essential considerations seem to be, first, that "living" forces are not spatially representable, and hence not mathematizable,[30] and, second, that living forces are not conserved over time.[31] They come into existence and pass away without further cause. Thus, to require that matter (the object of physics) be conserved and that it be spatially representable is to disallow the possibility of attributing "living" forces to matter. But, at the same time, this is to guaran-

[28] *MFNS*, 544.

[29] See the "Note" to Proposition 1 of the Metaphysical Foundations of Mechanics.

[30] *MFNS*, 543.

[31] *Ibid.*, p. 550.

tee the mathematizability of the matter concept, at least from this point of view,[32] and therefore to insure the "possibility" of a *proper* science of nature. It is in this way that the "first and second laws of mechanics" serve as presuppositions of physics. From the same point of view, the Cartesian concept of substance states the more general requirement that objects, insofar as they can be experienced by us, must be completely determinate.

The so-called "force of inertia" is for Kant the paradigm of a "living" force. In banning "living" forces from physics, he intends in the first place to eliminate the concept of inertial force from physics.[33] This elimination has three aspects. One, just noted, is that inertial force precludes the "possibility" of a mathematical physics.[34] Another aspect is that the concept of inertial force is empty. As Kant puts it in the Metaphysical Foundations of Mechanics: "Nothing but the opposite motion of another body can resist a motion, but this other's rest can in no way resist a motion. Here, then, inertia of matter, i.e., mere incapacity to move of itself, is not the cause of a resistance. A special and entirely peculiar force merely to resist, but without being able to move a body, would under the name of a force of inertia be a word without any meaning" (551). There is no more to the concept than what is already contained in the law of the equality of action and reaction.

[32] Clearly a problem remains concerning the extent to which the fundamental forces, attraction and repulsion, are mathematizable. Kant suggests that the fundamental forces are *determinate* in that each has a determinate degree (*MFNS*, 499; recall in this connection that the "Dynamics" is intended to parallel the Anticipations of Perception in the *Critique*) and in that each operates according to a mathematical law. But, as we saw earlier in this chapter, the fundamental forces are also inconstructible, i.e., according to Kant the operation of these forces cannot be represented spatially. Neither attraction nor repulsion, however, is a "living" force.

[33] It is possible, although I know of no direct evidence for it, that Kant was influenced by Hume's critique of the concept of *vis inertiae* in a footnote to Part I of the discussion "Of the idea of necessary connection" in the first *Enquiry*, p. 57n. of the Selby-Bigge edition of the *Enquiries* (Oxford: Clarendon Press, 1902).

[34] Inertial force is hylozoistic, a force whose occurrence is unpredictable, very much on the order of a human will.

The third aspect of Kant's banning of inertial force has to do with his realization, noted in chapter 4, that an object in uniform, rectilinear motion may indifferently be regarded as "in motion" or "at rest" (vis-à-vis a particular inertial frame). There is, therefore, no reason to introduce inertial force as the supposed cause of its continuing in motion.[35]

On the other hand, if rectilinear motion does not require the existence of sustaining causal forces, acceleration does. From a phoronomic or kinematic point of view, where attention is confined to the motion of individual objects represented as points in space, it is not possible to distinguish between the motion of an object and the (relative) space in which it is located. In dynamics, where impenetrability is chiefly at stake, an object is regarded not as in motion with respect to some space, but as the locus of central forces. Only in mechanics (as conceived by Kant) does it become possible to distinguish between objects at rest and in motion. According to the "second law of mechanics," a "motion insofar as it arises must have an external cause." An accelerated motion, in particular, requires a moving force. But since it is not possible to attribute this moving force to the space in which an object is located, we must attribute it to the object; i.e., in such circumstances we can say that the object *really* is in motion with respect to the surrounding space. This is an important consideration for two reasons. On the one hand, the reality of matter depends on it. Matter is characterized generally by Kant as the movable in space. It is an object of experience only to the extent that its motion can be discriminated, i.e., only to the extent that it has certain force-effects. But its motion can be discriminated only to the extent that it is not uniform. On the other hand, the reality of change depends on it. When Kant talks about change, primarily in connection with the measurement of time, he has change of spatial relations, or motion, in mind. With respect to uniform, rectilinear motion, as we have seen, there is no way to distinguish between the apparent and the real. Only acceleration, by way, once

[35] A point apparently not appreciated in *Thoughts on the True Estimation of Living Forces*.

again, of the "laws of mechanics," allows us to distinguish between apparent and real motion, and hence to distinguish between apparent and real change. Real changes, moreover, are required by the possibility of objective time-relations, as we shall see in the next chapter.

On my reading, then, the first and second Analogies of Experience, and in particular the first and second "laws of mechanics," serve as presuppositions of physics because they help to insure the mathematizability and the objectivity of the "objects" of physics—matter, force, and motion.

One more detail from the Metaphysical Foundations of Mechanics may serve both to clarify and to support this reading. It has to do with Kant's proposition that "the quantity of a matter can be estimated in comparison with every other matter only by its quantity of motion at a given velocity." In Definition 1 of *Principia*, Newton offers the following as a definition of "quantity of matter" or "mass." "The quantity of matter is the measure of the same, arising from its density and bulk,"[36] i.e., the product of volume and density. This explicit definition of quantity of matter is followed by an implicit definition in the commentary accompanying Definition 3 of *Principia*, in terms of inertial force, the measure of an object's resistance to change in its state of motion or rest.[37] The first definition has traditionally been much criticized, largely because of its apparent circularity.[38] More interesting for my purposes is Kant's criticism of the implicit definition of mass in terms of inertial force. We have already noticed his rejec-

[36] From the Cajori edition.

[37] See the discussion of Newton's views, and Kant's criticism of them, in Max Jammer, *Concepts of Mass* (New York: Harper Torchbooks, 1964), chapters 6-7. According to Jammer, pp. 82-83, "Kant concluded that vis inertiae . . . is fundamentally a superfluous and unnecessary concept." On my reading, Kant's claim is not just that the concept of inertial force is superfluous and unnecessary, but that it precludes the "possibility" of a mathematical physics.

[38] Since density would seem to be definable only as mass per unit volume. Note that in Book 3 of *Principia, The system of the World*, Proposition 6, Corollary 4, "bodies of the same density" are defined as those "whose inertia are in proportion to their bulks." See Jammer, p. 64.

tion of the concept of inertial force. It is clear from his discussion in the Metaphysical Foundations of Mechanics that a corresponding rejection of a definition of mass in terms of inertial force is tied to the fact that on such a definition "mass" is an *intensive* quantity.[39] Kant seems to think that quantity of matter must be extensive, since matter as substance is essentially spatial. But the deeper point is that matter as an object of the external senses must be an extensive magnitude. Otherwise it is indeterminate, hence not a possible object of experience, and unmathematizable, hence not a subject for scientific investigation. Kant proposes instead to define mass in terms of "quantity of motion" (momentum) at a given velocity.[40] On this definition, whatever other problems it might have, mass is an extensive magnitude. One might then go on to point out that the "three laws of mechanics" that follow this definition of quantity of matter elaborate different aspects of it. Thus, the "first law of mechanics" states the law of conservation of momentum, the "second law" states that velocity changes as a function of impressed force, and the "third law" simply explicates the concept of "momentum."

[39] *MFNS*, 541: ". . . it is clear that the quantity of substance in a matter must be estimated mechanically, i.e., by the quantity of the proper motion of the matter, and not dynamically, by the quantity of its original moving forces."

[40] I.e., mass = momentum/velocity.

Chapter 7: time and causality

IN the case of mathematical propositions, the main difficulty for a would-be Kantian is to show that they are synthetic. In the case of the Categories or the principles of the metaphysics of nature,[1] the difficulty is to show that they are *a priori*. For a standard criticism of Kant is that such propositions are neither "necessary" nor "presupposed." At best, they express certain methodological imperatives that have long guided the scientific enterprise. The criticism continues, moreover, that insofar as the propositions of mathematics, at least as concerns Euclidean geometry, or pure natural science are taken as synthetic, they are false, as developments in the history of science have shown.

But there is a more basic symmetry involved here. In both cases, the viability of a particular "reductionist" account is at stake. A received view of mathematical propositions turns in large part on an assumed reduction of mathematics to logic. The reduction, if successful, shows that mathematical propositions are analytic. I have tried to set out Kant's reasons for thinking that the reduction cannot be carried out successfully, hence that mathematical propositions are synthetic. A received view of the Categories, and more especially of the principles of the metaphysics of nature, turns on their assumed reduction to patterns of sense-experience—objects and causes are construed in terms of bundles and regular sequences of perceptions respectively—or, if one is not a phenomenalist, in terms of the ways in which the world appears

[1] I have rather blurred the distinction Kant wants to make between the Categories as *pure* and "*schematized*" and between the Categories (pure or schematized) as *concepts* and the Principles as *propositions*. Principles, as conditions of the possibility of experience, are judgments in which the Categories find necessary application. From the point of view of my exposition, the more important distinction is that between the "transcendental" Principles of the *Critique* and the "metaphysical" Laws of the *Metaphysical Foundations of Natural Science*.

to us. This time, the reduction shows that the propositions in question are *a posteriori*. I will now set out Kant's reasons for thinking, once again, that the reduction cannot be carried out, hence that these propositions are *a priori*.[2] In general, this comes to saying that objective time-determination is not possible on the basis of the contents of perception, what we are "given" in experience, alone; certain conceptual abilities on our part are also required. In particular, in the second Analogy, it comes to showing that the possibility of locating events vis-à-vis one another in time requires the application of the category of causality to our experience.

events and causes

The materials out of which Kant builds his argument are the concepts of an objective time-order, of an event, and of a cause. The Transcendental Deduction has as a corollary that the unity of consciousness is possible only on condition that the objects of experience are objectively ordered in time. The first Analogy, on the "Aristotelian" concept of substance, suggests a connection between the concept of an objective time-order and the concept of an event. It is the task of the second Analogy to complete the argument, by establishing a connection between the concept of an event and the concept of a cause.

Kant formulates his task as follows. We know from the argument of the first Analogy that the empirical, objective determination of time-relations must be by way of the perception of events and, further, that events must be analyzed as changes in at least relatively permanent, spatially extended, impenetrable objects. Under what conditions, then, can our perceptions be taken as perceptions of changes in objects?

It is clearly a necessary condition of a perception's being a

[2] Although I think the situation is very different here than with respect to Kant's philosophy of mathematics. Kant wants to defend a causal theory of time. But a causal theory of time is tied very closely to a particular development of physical theory, and physical theory has changed radically. There has been no comparable shift in mathematics, and as a result Kant's philosophy of mathematics is more defensible.

perception of objective change that the perception be of successive states of an object. But how can we distinguish within experience between a perception of successive states of an object as, to use Kant's example, in the case of a boat moving downstream, and a succession of perceptions of co-existing states of an object as, once again to use Kant's example, in the case of scanning a house from top to bottom? Both cases involve a succession of perceptions. Kant's answer is that in the case of objective change, the succession of perceptions is determinate, necessary, whereas in the case of perceiving a static state of affairs it is not. We distinguish between perception of events and perception of non-events in virtue of the fact that the successive order of perceptions in the case of event-perception is necessary. But the order of succession is necessary just in case the objective change is itself causally determined. Thus, perception of an event is in part perception of a causally determined succession of states of an object. The concept of an event connects with the concept of a cause in just this way.

With minor variations, this is how a great many commentators have summarized Kant's argument in the second Analogy. Thus, for a fairly recent example, Graham Bird:

"The analysis of an event has shown that the perception of different states in an object is not enough to discriminate events from non-events. What distinguishes one from the other is that events are regarded as ordered in a determinate way in time. What we mean by 'event' is such a determinate temporal order of two states in the same object. But the idea of a determinate order between two states presupposes that of a something which determines it; and this idea of a determinate or reason for such an order is that of a cause."[3]

Indeed, Kant himself seems to summarize his argument in the same terms in the paragraph added to the discussion on the second edition of the *Critique:*

"In order that . . . the *objective relation* of appearances that follow upon one another . . . be known as determined the re-

[3] *Kant's Theory of Knowledge*, p. 161.

lation between the two states must be so thought that it is thereby determined as necessary which of them must be placed before, and which of them after, and that they cannot be placed in the reverse relation. But the concept which carries with it a necessity of synthetic unity can only be a pure concept that lies in the understanding, not in perception; and in this case it is the concept of the *relation of cause and effect*, the former of which determines the latter in time, as its consequence . . ." (B234).[4]

But, set out just like this, the argument is often thought to involve two fallacies. Consideration of these alleged fallacies will bring us closer to the key premises of the argument and clear the way for our eventual reconstruction of it.

The first fallacy alleged is that the argument involves a non-sequitur.[5] Kant claims that we distinguish between perception of events and perception of non-events in virtue of the fact that the successive order of perceptions in the case of event-perception is necessary. Events are ordered pairs of states of an object. To say that the order of perceptions in the case of event perception is necessary is simply to say that another order of states of an object would constitute a *different* event.[6] As Bird puts it: "The necessity in such cases is the logical necessity that to apprehend a ship's sailing downstream is, necessarily, to apprehend an event in which the ship's position downstream followed its position upstream. The order of this event is a necessary order, not because it is impossible for ships to sail upstream, but because if the constituent states had been reversed the event apprehended would have been a different event. It would have been the event of a ship's sailing upstream."[7] But of course it does not follow from the fact that events are ordered pairs of objects that the order is caus-

[4] See also A192/B237-A194/B239.

[5] See P. F. Strawson, *The Bounds of Sense*, pp. 137ff., and W. A. Suchting, "Kant's Second Analogy of Experience," *Kant-Studien* (1967), pp. 162-163.

[6] Discounting differential time-lags, etc., if state A precedes state B, then perception of A will precede perception of B. Any other order of perception would indicate perception of a different event, if in fact an event is being perceived.

[7] *Kant's Theory of Knowledge*, p. 155.

ally determined. Only by punning on the word "necessity" could one get from the logically necessary order of event-states to their causally necessary, i.e., determined, order. Yet this is just what Kant seems to do.

The second alleged fallacy is that the argument involves a circularity.[8] Kant claims that events are temporally ordered pairs of states of an object. ". . . the relation between the two states must be so thought that it is thereby determined as necessary which of them must be placed before, and which of them after, and that they cannot be placed in the reverse relation." What determines the temporal order is the relation of cause and effect. To say that *B* follows *A* is to say that *A* is the cause of *B* or is simultaneous with the cause of *B*: ". . . it is the concept of the relation of cause and effect, the former of which determines the latter in time, as its consequence . . ." which "carries with it" the requisite "necessity of synthetic unity." But this is circular. For it is part of the concept of *A* causing *B* that *B* follows *A* in time.[9] More generally, Kant wants to determine objective time-relations by way of our perception of events. But events are distinguished from non-events in terms of their causally ordered character; that is, the event-states that compose them are causally ordered. But to say that two event-states are causally ordered is to say, among other things, that one precedes the other in time. The concept of cause that was presumably presupposed by objective time-determination itself presupposes that time can be objectively ordered.

[8] See Jeffrie G. Murphy, "Kant's Second Analogy as an Answer to Hume," *Ratio* (1969), pp. 75-78, and Suchting, "Kant's Second Analogy of Experience," pp. 367-369. Suchting does not think that Kant's argument is circular, only a mistaken interpretation of it. In one form or another, the objection seems to originate with Schopenhauer.

[9] As Kant seems to admit at A243/B301: "If I omit from the concept of cause the time in which something follows upon something else in conformity with a rule, I should find in the pure category nothing further than that there is something from which we can conclude to the existence of something else. In that case not only would we be unable to distinguish cause and effect from one another, but since the power to draw such inferences requires conditions of which I know nothing, the concept would yield no indication how it applies to any object."

the causal theory of time

In my view, the argument of the second Analogy should be set out in the following way. The first premise is that we perceive that "appearances follow one another; that is, that there is a state of things at one time the opposite of which was in the preceding time" (B233). The second premise is that time-order, one event or event-state taking place earlier (later) than another, cannot be determined with respect to time itself. Kant puts this in terms of the imperceptibility of time, as at B233: "For time cannot be perceived in itself, and what precedes and what follows cannot, therefore, by relation to it, be empirically determined in the object." What is needed is an empirical criterion of time-order. The third premise is that the mere order of our perceptions does not provide an adequate empirical criterion. An adequate criterion must be objective as well as merely empirical, and "the objective relation of appearances that follow upon one another is not to be determined through mere perception" (B234). The mere order of our perceptions does not allow us to distinguish within our experience between changes taking place in objects in time, that is, events, and static states of affairs. In both cases, our perceptions might very well be successive. Further, it is a demand of the Transcendental Deduction that we be able to distinguish within our experience between a mere succession of perceptions and perceived changes in objects, between our experiences and a world that is the object of such experiences. The fourth premise is that not the *mere* order, but the *necessary* order of our perceptions, does provide an adequate criterion, both empirical and objective, of time-order. The criterion is empirical because it is phrased in terms of our perceptual experience. The criterion is objective because it allows us to distinguish between a perception of succession and a succession of perceptions. But to say that the order of our perceptions is necessary is to say that the order of change in the object perceived is causally determined. Thus, the fact that some of our perceptions are perceptions of events entails that changes in objects be causally determined. In this sense, causal determination is a necessary condition of experience. Or, more generally, our experience must be conceptualized in accordance with the category of causality if our experience is to be of objects ordered in time.

This way of setting out the argument, really no more than a paraphrase of Kant's summary in the second edition of the *Critique*, bears obvious similarities to the way in which the argument was set out earlier, and on the face of it does not avoid the two fallacies already considered. I put it in this form mainly because I want to emphasize the notion of an objective, empirical criterion of time-order. On my reading, what is central to the argument is the claim that causal order provides such a criterion of time-order.

Let us try to make this more precise.[10] We want a criterion for determining when one event or event-state (since an event, of the simple sort we have considered so far, is made up of successive states of an object) E_1 is earlier (later) than another event or event-state D_2 that does not itself appeal to temporal order. On our reconstruction of Kant's position, we

[10] In what follows I am indebted to Hans Reichenbach's discussion in #21 of *The Philosophy of Space and Time* and to Adolf Grünbaum's discussion in chapters 7 and 8 of *Philosophical Problems of Space and Time*. See also H. Scholz, "Eine Topologie der Zeit im Kantischen Sinne," *Dialectica*, IX (1955), and H. Mehlberg, "Physical Laws and Time's Arrow," in H. Feigl and G. Maxwell, eds., *Current Issues in the Philosophy of Science* (New York: Holt, Rinehart, & Winston, 1961).

[11] Reichenbach does not include the concept of simultaneity in his statement of the criterion since he takes it to be definable in terms of the "later than" relation. I take it that the purpose of Kant's *third* Analogy is to provide a physical basis for simultaneity. See the *Prolegomena*, #25: "In respect of the relation of appearances, solely with regard to their existence, the determination of this relation is not mathematical but dynamic, and can never be objectively valid and fit for experience if it does not stand under principles *a priori* which first make possible knowledge by experience in respect of it. Hence appearances must be subsumed under the concept of substance, which, as a concept of the thing itself, lies at the ground of all determination of existence; or secondly, in the case of a succession in time among the appearances, i.e. and event, under the concept of an effect in reference to a cause; or in the case of contemporaneity which is to be known objectively, i.e. through a judgment of experience, under the concept of community (reciprocity)." It is questionable whether in fact Kant succeeded in providing an adequate physical basis for simultaneity and, indeed, whether an adequate basis can be given in the context of classical physics.

might, following Reichenbach, put the criterion as follows:

> *CTO: If E_2 is the effect of E_1, or is the effect of an event simultaneous with E_1,*[11] *then E_2 is later than E_1.*[12]

Now how, in the first place, does this criterion avoid the charge of circularity? For doesn't saying that E_1 is a cause of E_2 or that E_2 is an effect of E_1 involve prior appeal to the temporal order of E_1 and E_2? Reichenbach's response in *The Philosophy of Space and Time*[13] is roughly as follows. The cause-effect relation serves not only to connect pairs of events, but to order them asymmetrically. It thus defines an order. But no reference to time need be made in determining this order in particular cases. For saying that E_1 is a cause of E_2 or that E_2 is an effect of E_1 need only involve appeal to the observation that small variations in E_1 are associated with small variations in E_2 whereas small variations in E_2 are not associated with corresponding variations in E_1. Reichenbach gives two examples:

(1) "We send a light ray from A to B. If we hold a red glass in the path of the light at A, the light will also be red at B. If we hold the red glass in the path of the light at B, it will *not* be colored at A."

(2) "We throw a stone from A to B. If we mark the stone with a piece of chalk at A, it will carry the same mark when it arrives at B (event E_2). If we mark the stone only at its arrival at B, then the stone leaving A (event E_1) has no mark."

In both examples, we seem to be able to order E_1, E_2 independently of temporal considerations, by using this so-called "mark principle." Therefore, according to Reichenbach, causal order can be used, non-circularly, as a criterion of time order.

Unfortunately, the "mark principle" is not as helpful as Reichenbach suggests, for it ultimately depends for its appli-

[12] We can forget about whether E_1 is only a *partial* cause of E_2 or E_2 only a *partial* effect of E_1.

[13] He modified his position in various important ways in later publications, e.g., in *The Direction of Time* (Berkeley: University of California Press, 1956).

cation on a prior appeal to temporal order, and hence it cannot be used to provide a non-circular criterion for temporal order between pairs of events.[14] Take example (2) as a case in point. Suppose that we throw the stone from *A* to *B* a number of times. It is possible that the stone arrives at *B* unmarked and then is returned to *A*, where it is marked and then thrown to *B* again, in which case we have the sequence of events E_2, E_1, E_2. If we consider the first two events in this sequence as a pair, then it is not true that a small variation in E_1 is accompanied by a small variation in E_2. The "mark principle" works only if we consider E_1 and E_2 as an event-pair. But the only way of isolating the appropriate event-pairs, apparently, is to take them in the sequence E_1, E_2 and this evidently involves appealing to temporal order.

Kant's position differs importantly from Reichenbach's in that he allows himself to use, perhaps too uncritically, the much richer concept of a causal law. For a reconstructed Kant, to say that E_1 causes E_2 is to say that there is a law or, in Kant's terminology, rule which taken together with a description of E_1 allows us to infer a description of E_2.[15] And, at least at first glance, this way of characterizing the cause-effect relation is independent of temporal considerations.[16] Laws allow us to order events causally and the causal order of events allows us to order them in time empirically and objectively. Which is to say that if this way of characterizing the cause-effect relation is independent of temporal considerations, the

[14] See the criticisms by Grünbaum and Mehlberg in the works mentioned in footnote 10.

[15] This is obviously a very simplified picture. For ordinarily E_1 is only one among a number of initial and boundary conditions which taken together with a law or laws allows us to infer a description of E_2. Notice that in such a picture E_1 and E_2 are usually thought of as event-states rather than as events. In mechanics, for example, the deduction goes from the initial mechanical state of a system via Newton's laws to another mechanical state of the system.

[16] For a second glance see van Fraassen's criticisms of classical causal theories of time, *An Introduction to the Philosophy of Time and Space*, Chapter II, section 3. Van Fraassen concentrates on the problem of giving a physical basis for simultaneity and on specifying the concept of a mechanical state without smuggling in temporal concepts.

concept of a law does not itself contain the concept of temporal order.

Kant makes the connection between causes and laws explicit at a number of different points. For example, as he says at A539/B567 of the *Critique*: "Every efficient cause must have a character, that is, a law of its causality, without which it would not be a cause." Then again, in more detail, at A549/B576: "For every cause presupposes a rule according to which certain appearances follow as effects; and every rule requires uniformity in the effects. This uniformity is, indeed, that upon which the concept of cause . . . is based, and so far as it must be exhibited by mere appearances may be named the empirical character of the cause."

In the second place, Kant's characterization of causality implies that the relation between descriptions of events causally related is deductive. Consider, for example, the formulation of the "law of causality" given in the first edition of the *Critique*: everything that happens, that begins to be, presupposes something upon which it follows according to a rule." The "follows" here is, I think, to be taken first not in a temporal but in a logical sense. If an event takes place, then it is caused, but this means that a description of it can be deduced from a description of another event together with a rule or law that functions as a major premise in the deduction. Compare the passage at A193/B238, where Kant says "But in the perception of an event there is always a rule that makes the order in which the perceptions (in the apprehension of this appearance) follow upon one another a *necessary* order." The necessity, once again, is a function of the fact that, given the rule and an initial event or perception of an event, other events or perceptions of events follow, logically. The possibility of reading these passages in this way is very much reinforced by recalling that the pure concept corresponding to the category of causality is that of ground and consequent. Causality involves a logical relation.

But if rules or, in our terminology, laws causally order events, or rather provide a criterion by which they may be causally ordered, they order events temporally as well. For it follows from the fact that one event causes another that the first is earlier than the second—by the CTO criterion. So the "follows" in the passages at A189, at B193/B238, and also at A194/B239, may be taken in a temporal sense as well. The important point is that it is the causal order that makes the temporal order possible in the first place, and that the causal order is made possible only by assuming that the succession of states in an object is rule-governed.[17]

I have been arguing that Kant subscribed to what is usually called a causal theory of time: temporal order is determined by causal order. I want now to defend this claim against certain objections that might be made against it.[18] Most of these objections point back toward the charge of circularity.

According to the CTO criterion, if one event causes another, then it precedes it. Further, one event causes another just in case from a description of the first event and a given rule or law we can derive a description of the second event. The laws of Newtonian mechanics have this peculiarity, however, that their temporal parameters are reversible. Given a description of a state of a mechanical system, one can derive via these laws a description of any other state of the system, past as well as future. In other words, these laws, which allow for the reversibility of the mechanical processes they describe, do not determine a unique ordering or "direction" of time. Depending on which we take as premise and which as conclusion, one event will be earlier than or later than another; we cannot say, *tout court*, that one is definitely earlier or definitely

[17] See, for example, A199/B244–A200/B245: "Understanding is required for all experience and for its possibility. Its primary contribution does not consist in making the representation of objects distinct, but in making the representation of an object possible at all. This it does by carrying the time-order over into the appearances and their existence. For to each of them, (viewed) as (a) consequent, it assigns, through relation to the preceding appearances, a position determined *a priori* in time." See also A788/B816.

[18] I.e., I want to defend the claim that Kant subscribed to a causal theory of time, not the claim that a causal theory of time is correct. Grünbaum and van Fraassen have provided sophisticated defenses of the causal theory itself.

later than another.[19] Reversible processes do not establish an asymmetric causal relation, hence do not establish an asymmetric temporal relation.

Much of the time it appears as though Kant simply assumes the irreversibility of the rules to which he appeals, as at A193/B238-A194/B239:

"In conformity with such a rule there must lie in that which precedes an event the condition of a rule according to which this event invariably and necessarily follows. I cannot reverse this order, proceeding back from the event to determine through apprehension that which precedes. For appearance never goes back from the succeeding to the preceding point of time, though it does indeed stand in relation to *some* preceding point of time. The advance, on the other hand, from a given time to the determinate time that follows is a necessary advance."

Perhaps Kant thinks that Newton's laws, for him the paradigm of causal "rules," are irreversible.[20] On the other hand, we might interpret his position as *requiring* that there be irreversible laws. But this seems unduly a prioristic, even for Kant.[21] Or, as a third possibility, Kant simply wants to claim that, although not precluded by law, the temporal inverse of physical processes never in fact occurs.[22] From this point of

[19] In a more technical idiom, time defined in terms of reversible laws is *isotropic*, rather than *anistropic*.

[20] Lagrange seems to have been the first to emphasize "that the time variable of rational mechanics based on Newton's laws of motion does not point in a unique direction and that, in principle, all motions and dynamical processes subject to these laws are reversible, just like the axes of geometrical systems." G. J. Whitrow, *The National Philosophy of Time* (London: Thomas Nelson & Sons, Ltd., 1961), p. 4.

[21] Perhaps there are such laws, possibly the entropy law of classical thermodynamics, which provide for an asymmetric causal relation and hence a physical basis for the anistropy of time. But the existence of such laws seems very much a contingent matter. Surely it would be inappropriate to argue that the second law of thermodynamics "vindicates" Kant.

[22] A "weak" sense in which a process might be claimed to be "irreversible" has been sketched by Grünbaum: "the *temporal inverse* of the process never (or hardly ever) occurs with increasing time for the following reason: Certain boundary or initial conditions obtaining in the universe independ-

view, we are to read "appearance never goes back from the succeeding to the preceding point of time" in the above passage as a statement of fact.

In any case, I think we can redeploy Kant's position so that it is at least consistent with it that the laws that describe natural processes are reversible. As Grünbaum has maintained,[23] *de facto* reversible processes define (under suitable boundary conditions) a temporal order of betweenness among events. In that case, we can introduce a serial relation between the events, by assigning larger and smaller real numbers to arbitrarily chosen reference points in this order. Of course, the serial relation would not allow us to attribute a "direction" to the "flow" of time (since the assignment of real numbers is arbitrary), but it would allow us to order events temporally. At the same time, insofar as the serial relation was *introduced*, it would be extrinsic rather than intrinsic. And this fact could be used, against the background of the discussion in chapter 4, to provide more support for the Kantian claim that in an important respect time is "subjective."[24]

There are a few more loose ends. It is sometimes suggested that attributing a causal theory to Kant, as I have done, must be mistaken because Leibniz held a causal theory of time and Kant criticizes Leibniz's position on time. But, as I have tried

---

ently of any law (or laws) combine with a relevant law (or laws) to render the temporal inverse *de facto* nonexistent or unreversed, although no law or combination of laws itself disallows that inverse process." "The Anistropy of Time," in T. Gold, ed., *The Nature of Time* (Ithaca: Cornell University Press, 1967), p. 160.

[23] See Grünbaum, *Philosophical Problems of Space and Time*, pp. 216ff.

[24] In *The A Priori in Physical Theory*, p. 70n., Arthur Pap says that "This intimate connection between time series and causal series is an immediate consequence of the *relative* view of time adopted by Kant, viz., the view of time as an order of events. Thus, as far as in mechanics causal sequences are reversible, i.e., the past can be inferred, logically reconstructed, from the present, time too is said to be reversible—a mode of speech that sounds paradoxical only if we think of absolute time." Pap, however, seems to run two different considerations together. For a relative (i.e., relational) view of time as the order of events is independent of any assumptions concerning the reversibility or irreversibility of the basic laws of physics.

to show in earlier chapters, Kant accepted Leibniz's view as far as *determinate* time-relations, within experience, were involved. See, for instance, the note at A452/B480 of the *Critique*—"Time, as the formal condition of the possibility of changes, is indeed prior to them; subjectively, however, in actual consciousness, the representation of time, like every other, is given only in connection with perceptions"—and the letter to Reinhold of May 19, 1789, where Kant refers to "Leibniz's correct opinion" about the *concrete* conception of time.

Finally, how can attributing a causal theory of time to Kant be reconciled with the passage already quoted at A243/B301, or the passage at A203/B249: "The sequence in time is thus the sole empirical criterion of an effect in its relation to the causality of the cause which precedes it?" It is difficult to know what to say about the former passage, for in it Kant appears to be running several different points together. As for the latter passage, Kant simply puts his point in a misleading way. He asks how we distinguish between cause and effect when cause and effect are simultaneous.[25] His answer is that we "distinguish the two through the time-relation of their dynamical connection." But these words are not to be taken at face value, as the accompanying example reveals: "For if I lay the ball on the cushion, a hollow follows upon the previous flat smooth shape; but if (for any reason) there previously exists a hollow in the cushion, a leaden ball does not follow upon it." We do not distinguish cause from effect here on the basis of the fact that putting the ball on the cushion temporally precedes the hollow forming, for *ex hypothesi* the two events are simultaneous.[26] Rather, we distinguish the two on the basis of their *dynamical* connection. What is their "dynam-

[25] Granting for the sake of argument the possibility of what Hume, perhaps unbeknowst to Kant, denies.

[26] Kant's admission that causes are sometimes simultaneous with effects is surely incompatible with the claim, taken at face value, that "The sequence in time is thus the sole empirical criterion of an effect in its relation to the causality of the cause which precedes it." Far from undermining attributing a causal theory of time to Kant, as some commentators maintain, I would suggest that this assertion supports doing so.

ical" connection? Simply, that if I were to put a ball on the cushion, it would follow (by way of a causal law, not temporally) that a hollow would form. Whereas if I were to form a hollow, it would not follow (i.e., there is no rule to this effect) that a ball had been put on the cushion. If in doing one thing, I bring about another, then we say that the first is the cause of the second.[27] But this is just what it means to say that a law "connects" them. This is also what it means to say that the causal relation is asymmetrical, hence, at least in Kant's view, this is the ultimate asymmetry of the temporal relation.

I have claimed that Kant's argument, correctly understood, does not involve a circularity. Does it involve a pun on "necessity," illicitly shifting from logical necessity to causal necessity? If my reconstruction is correct, it does not. It is true that for Kant the order of event-states constituting an event is logically necessary. But the logical necessity involved has nothing to do with the fact that if the order or event-states were changed they would constitute a different event. Rather, it is simply that from a description of the first event-state we can, with the help of the appropriate "rule," derive a description of the second event-state, and so on. It is logically necessary that given "rules" or natural laws, the first event-state in a series will determine the successive members of the series—just as we can derive statements describing the position of a boat at one time from other statements describing the position of the boat at another time and the laws of motion. But this logical necessity is perfectly compatible with the empirical character of the series. In Kant's view, to say that one event causes another is just to say that there is this type of logical connection between them. There are philosophical grounds on which this view may be challenged. I noted at the outset that Kant uses the concept of a "rule" in what is perhaps too uncritical a way.[28] But questionable or not, it is a view which does not simply pun on "necessity."

[27] See G. H. von Wright, *Explanation and Understanding* (Ithaca: Cornell University Press, 1971), pp. 74ff.

[28] I will say something more about "rules" in the next section of this chapter and in the chapter following.

causality and objectivity

This outline of Kant's argument can be made clearer by considering it in two slightly different contexts. One of these contexts has to do with the second Analogy as a "reply to Hume." Reams have been written on this subject. To put it bluntly and briefly, not very much of what has been written is persuasive. On my account, one aspect of Kant's "reply to Hume" is as follows.[29] The cause-effect relation provides the only empirical, objective criterion of time-order. Such an order depends on our assuming in advance that rules or laws connecting events or event-stages can be found. Therefore, this general assumption—that "all alterations take place in conformity with the law of the connection of cause and effect"—cannot be reduced to or derived from patterns of sense experiences or the repeated observations of events. An event $E1$ is earlier than an event $E2$ just in case $E1$ causes, or is simultaneous with the cause of, $E2$. But $E1$, or an event simultaneous with it, caused $E2$ just in case there is a law connecting them. If Hume is right, we cannot know in advance that there are laws connecting them. It is a merely contingent matter, then, given the causal theory of time, whether any two events can be located vis-à-vis one another temporally. But this, as the Transcendental Deduction has shown, is incompatible with the unity of consciousness, for the possibility of attributing experiences to myself as their subject depends on my being able to locate them in time.

Notice that Kant does not disagree with Hume about the character of the individual empirical laws that in fact govern our experience. These are contingent;[30] they are discovered as regularities in our experience. Hume's mistake, according to Kant at A766/B794 of the *Critique*, lay in "inferring from the contingency of our determination *in accordance with the law* (i.e., of causality) the contingency of the law itself." It is necessary that our experience be law-governed, for the rea-

[29] Another aspect of the "reply to Hume" will be developed in the next chapter.

[30] As we shall see in a page or two, they are not "merely" contingent.

sons I have set out. But it does not follow from that fact that the particular laws that govern experience must be, in the same strong sense, necessary. According to Kant, Hume's mistake was in thinking that since the entailment held, and the consequent was false, the antecedent—that all our experience must be law-governed—must also be false.[31]

The other context in which my account of the second Analogy might be put in the attempt to clarify it has to do with what might be called the *compulsive* character of reality. Kant tends to identify what is objective with what is in some sense necessary and what is necessary with what is compelled. For example, at A1916/B241-A197/B242:

"We have, then, to show, in the case under consideration, that we never, even in experience, ascribe succession (that is, the happening of some event which previously did not exist) to the object, and so distinguish it from subjective sequence in our apprehension, except when there is an underlying rule which compels us to observe this order of perceptions rather than any other; nay, that this compulsion is what first makes possible the representation of succession in the object."

The distinction between our perception of the successive states of a ship as it moves downstream and our successive perception of co-existent states of a house or, more generally, between our perception of events and our perception of non-events can be put by saying that the sequence of perceptions in the former case is *compelled.* What this comes to, once again, is that the sequence is law-governed. Or, to put it in a slightly different way, given the position of the boat at one moment in the stream, its position and velocity, we can *predict* its position at any other moment. For to predict position is simply to derive a statement describing it from other statements, at least one of which is a rule or law. If the prediction turns out false, what is shown is that we do not as yet have the appropriate rule. On the other hand, if we do have the appro-

[31] It does not matter for my purposes whether in fact Hume was guilty of this mistake. More formally, it comes to thinking that

(x) [(Event$_x \rightarrow (Ey)$ (Necessarily (y caused x))] follows from

Necessarily [(x) (Event$_x \rightarrow (Ey)$ (y caused x))].

priate rule, then we must see (other things being equal) what we have predicted. But in the case of scanning a house, given one perception, say of the top floor, and no additional information,[32] then there is no way to make a parallel prediction about what will be seen next. The content of our perception is not compelled as it is in the case of watching a boat moving downstream. But to say that it is "compelled" is to say that it is a perception of objective change.

The same sort of point is at stake, I believe, in the distinction Kant makes in the *Prolegomena* between judgments of perception and judgments of experience. The former are merely subjectively valid, the latter have objective validity. "That the room is warm, sugar is sweet, wormwood is nasty, are merely subjectively valid judgments . . . when I say air is elastic, this judgment is at first only a judgment of perception. I only refer two sensations in my sense to one another. But if I would have called it a judgment of experience, I demand that this connection shall stand under a condition which makes it universally valid. I require that I and everybody must always and necessarily conjoin the same perceptions under the same circumstances" (*Prolegomena*, #19).

Kant is aware that his account is not overly clear and he tries an "easier example": "When the sun shines on the stone, it grows warm. This judgment is a mere judgment of perception and contains no necessity, no matter how often I and others may have perceived this: the perceptions are only usually found conjoined in this way. But if I say: the sun *warms* the stone, the concept of the understanding of cause is added to the perception and connects the warmth *necessarily* with the concept of sunshine. The synthetic judgment becomes necessarily valid, consequently objective and is converted from a perception into experience."

In sum, judgments of experience are objectively valid, the concepts in them are necessarily connected, and the concept

[32] There are, of course, laws connecting bodily movements, etc., with what is perceived, and on the basis of which we could predict what parts of the house would next be seen, but to take account of these movements, etc., would be to bring in "additional information."

of a cause has been "added" to them. At least three things here are puzzling. Why should "the sun warms the stone" be objectively valid and "when the sun shines on the stone, it grows warm" not; how can concepts be necessarily connected in what is clearly a synthetic *a posteriori* judgment; and what does it mean to say that the concept of a cause has been "added" or that judgments of perception are "converted" into judgments of experience.?

These are very difficult questions. In fact, Kant's attempt to work out an answer to them carries through the Appendix to the Transcendental Dialectic ("The Regulative Employment of the Ideas of Pure Reason") and into the *Critique of Judgment* (especially the Introduction). But I think we can clarify, in a preliminary way, both the distinction Kant has in mind and the sense of "necessary" that seems to be involved.[33] In the first place, the distinction between judgments of experience and judgments of perception corresponds, in a rough way, to the contemporary distinction between "laws" and "accidental generalizations." Judgments of perception, we could say, are instances of accidental generalizations; judgments of perception are instances of laws. This leaves us, of course, with the problem of distinguishing between accidental generalizations and laws,[34] but it does indicate that whatever "necessity" judgments of experience may be said to have does not derive from the degree to which they have been confirmed, for laws and accidental generalizations alike describe or express uniformities in our observational experience. It also indicates that judgments of experience are potentially predictive in character, whereas judgments of perception are not. In the second place, it is clear that whatever sense of "necessity" attaches to judgments of experience, it cannot be the same sense of "necessity" that attaches to the Principles of Pure Reason.[35]

[33] I am not going to try, in any detail, to go over terrain already brilliantly explored by Gerd Buchdahl in *Metaphysics and the Philosophy of Science*, chapter VIII.

[34] About which I will have more to say in the next chapter.

[35] Although, as is so often the case with Kant, one can see an analogy or rough parallel.

The form of experience in general is to be distinguished from the content of particular empirical laws. Moreover, the second Analogy, which necessitates the existence of "rules," does not require that these "rules" are themselves any more than contingent propositions. In the third place, Kant's language—that judgments of experience are "subsumed under" the category of causality—provides us with a clue that it is tempting to develop as follows. To say that the concepts in a judgment of experience are "necessarily conjoined" is not to say that one is "contained" in the other. Rather, it is the case that the concepts are conjoined in a "rule" or law in such a way that with respect to a particular judgment of experience the second concept can be derived from the first via a "rule" or law. By contrast, in judgments of perception there is no corresponding "covering" law. Nothing is "done" to a judgment of perception to "convert" it into a judgment of experience, except to claim (or to show) that it instantiates a law. But, given Kant's analysis of the concept, to say that it instantiates a law is simply to subsume it under the category of causality.

But this is only a preliminary, far from satisfactory, clarification of the issues. In particular, it does not suffice to say that judgments of experience but not judgments of perception are "covered" by laws.[36] On the one hand, we would still be left with the problem of "laws," of explaining the sense in which *they* are necessary, for, if some propositions are derivative, some must be underived. On the other hand, Kant's discussion of just this point in the Introduction to the *Critique of Judgment* both implies that the above account is mistaken as a reconstruction of his position and indicates yet another sense in which empirical laws may be said to be necessary:

[36] As Buchdahl points out, some philosophers have wanted to make the distinction between laws and accidental generalizations in just this way. Thus R. B. Braithwaite (developing the views of William Kneale) in *Scientific Explanation* (Cambridge: Cambridge University Press, 1955), p. 304: "The blackness of all ravens is surely 'accidental' if no reason can be given for such blackness; and this is equivalent to saying that there is no established scientific system in which the generalization appears as a sequence."

". . . the forms of nature are so manifold, and there are so many modifications of the universal transcendental natural concepts left undetermined by the laws given, *a priori*, by the pure understanding—because these only concern the possibility of a nature in general (as an object of sense)—that there must be laws for these also. These, as empirical, may be contingent from the point of view of *our* understanding; and yet, if they are to be called laws (as the concept of a nature requires), they must be regarded as necessary in virtue of a principle of the unity of the manifold, though it be unknown to us."[37]

The key phrase here, clearly, is "must be regarded as necessary in virtue of a principle of the unity of the manifold. . . ." At once this precludes "must be regarded as necessary in virtue of their derivability from other laws" (although derivability relations are one way in which the unity of a system of laws is expressed). More importantly, it suggests, first, that it is not that laws are necessary but that they must be *regarded* as necessary, and, second, that laws, hence also judgments of experience, must be regarded as necessary insofar as they constitute the representation of the systematic unity of nature that, Kant goes on to tell us in the Introduction to the third *Critique*, is required by reason in its regulative aspect. It is only insofar as we can "think" laws as "necessary," i.e., as part of a general plan (which we, in fact, do not and cannot know) in terms of which all natural phenomena are ultimately intelligible.[38] To talk about the "unity" of nature in this connection is simply to talk about natural phenomena *as though* they *all* had some inner purpose.

"weak" and "strong" causality

In our discussion of the first Analogy in the last chapter, I raised the question whether the strong claim Kant wanted to make or only something weaker followed. Now we have to

[37] From the Bernard translation, p. 16.

[38] In that sense, presumably, in which a human action is "intelligible" (or "rational") once we both see and appreciate the purpose or motive for which it was done.

consider what relation the conclusion of the second Analogy—"All alterations take place in conformity with the law of cause and effect"—bears on the account of Kant's argument I have given. Is it possible, as was the case with the first Analogy, that only a "weaker" conclusion follows, that for the objective ordering of time it is not required that *every* event be construed as having a cause but only a good number of events?

Strawson, for one, claims that only a "weaker" conclusion follows: "Kant argued . . . for the conclusion that there existed strictly sufficient conditions for absolutely every change that we can take cognizance of. Of course we cannot regard any such absolute conclusion as established by the considerations just put forward. We do not have to suppose that explanatory conditions, fully stated, of every change or absence of change, must be strictly sufficient conditions. We do not have to suppose that there must always be an explanatory condition if only we could find it. We could accommodate some inexplicable objective change, and some more exceptions to our law-like expectations, without damage to the necessary but loosely woven mesh of our concepts of the objective."[39]

I want to make two comments about this claim. The first comment concerns the assertion that "We do not have to suppose that explanatory conditions, fully stated, of every change or absence of change must be strictly sufficient conditions." I take this to be an assertion about the meaning of "explanatory conditions." Strawson gives no argument, but presumably he holds that explanatory conditions do not always have to be construed as sufficient conditions. Perhaps he is right about this. If the explanatory conditions are not sufficient, however, it is difficult to see in what sense they could be *determining* conditions.

The second comment concerns the assertion that "we could accommodate some inexplicable change." What does this mean? Presumably that some inexplicable change is compati-

[39] *The Bounds of Sense*, p. 146.

ble with an objective ordering of events in time. But, given Kant's argument in the second Analogy, to admit some inexplicable change is to give up the possibility of locating *all* events vis-à-vis one another in time. But since having a definite space-time location is at least a necessary condition for being *real*, we should have to say that such events are unreal or revise our concept of reality. In either case, some very deep conceptual revisions would seem to be forced on us.

Chapter 8: the problem of induction and its "solution"

THE celebrated "reply to Hume" assumes a variety of forms. Many commentators, for example, have taken the second Analogy to be Kant's solution to the problem of induction.[1] It is. But not for the reasons, or in the ways, usually suggested.

Hume's problem

As Hume develops it, the "problem" of induction is intimately connected with his analysis of causality.[2] On that analysis, to say that one event causes another is to say that events of the first kind are followed by events of the second kind; that is, there is a law or inductive generalization "covering" them. Hence justification of particular causal claims involves justifying the claim that the appropriate laws or generalizations obtain. But how are we to justify this latter claim? Not deductively, for the falsity of a (presumed) law or inductive generalization, as shown by some future observation, is compatible with the course of our past experience. Not inductively either, except on pain of circularity. But if laws or inductive generalizations cannot be justified either deductively or inductively, Hume concludes, then they cannot be justified at all. Induction has no justification, although it can in some sense be "explained," by referring to habit and custom.

[1] Significantly, Kant nowhere mentions a "problem" of induction, although an extensive discussion of inductive methods is included in his lectures on logic. On the other hand, he is very much concerned with the problem of law-likeness. I intend to deal only with several, very general, aspects of his discussion, again referring the reader to Buchdahl's *Metaphysics and the Philosophy of Science*, on whose detailed discussion of the problem as it is handled in the Appendix to the Transcendental Dialectic and the *Critique of Judgment* I have not been able to improve.

[2] This characterization of Hume's position, and a great deal of what follows, I owe to Donald Davidson, although I certainly do not want to hold him responsible for my elaboration of it.

There are those who contend that Kant tried to reply to Hume in the second Analogy by proving a "law of causality" or "principle of induction" that would, in fact, guarantee particular generalizations or laws.[3] Thus Sir Karl Popper: "Kant tried to force his way out of the difficulty by taking the principle of induction (which he formulated as the 'principle of universal causation') to be '*a priori*' valid."[4] There are, however, at least two objections to attributing this "solution" to Kant. In the first place, the "principle of induction"—the course of nature continues always the same—is not implied by the conclusion of the second Analogy—everything that happens, that begins to be, presupposes something upon which it follows according to a rule; nor is it easy to see how the conclusion of the second Analogy could be used to guarantee particular inductive inferences.[5] In the second place, a "principle of induction" guarantees a more than contingent status for the laws or generalizations that fall under it. For if the evidence for a law is that it has always been that way in the past, and if the "principle of induction" is to the effect that if it has always been that way in the past then it will continue to be that way in the future, then the law is obviously immune to recalcitrant observations. In other words, a "principle of induction" provides a deductive justification of particular laws or inductive generalizations. But, as I have insisted repeatedly, Kant denies that such a deductive justification is possible; laws are never more than contingent, relative to the evidence we have accumulated for them.

In my view, Kant at least implicitly accepts Hume's argument: induction cannot be justified either inductively or de-

[3] Hume's statement of such a principle is as good as any: "*that instances, of which we have had no experience, must resemble those, of which we have had experience, and that the course of nature continues always the same.*" *Treatise of Human Nature*, Book I, Part III, section VI (Selby-Bigge edition, p. 89).

[4] *The Logic of Scientific Discovery* (London: Hutchinson, 1959), p. 29. Popper adds that he does not think that Kant's "ingenious attempt to provide an *a priori* justification for synthetic statements was successful."

[5] See Wesley Salmon, "The Foundations of Scientific Inference," in *Mind and Cosmos*, R. Colodny, ed. (Pittsburgh: University of Pittsburgh Press, 1966), p. 177.

ductively. His "solution" to the problem of induction does not consist in trying to find such a justification.[6]

"grue"-type predicates

Let us grant that Hume has shown that induction cannot be "justified." We might then turn our attention, as Nelson Goodman does in *Fact, Fiction, and Forecast*, to a *description* of sound inductive procedures. Our problem is no longer to justify induction, but to say when a particular generalization is capable of being confirmed by its instances. To put it in a slightly different way, recalling the closing pages of the last chapter, our problem is to distinguish between "law-like" and "accidental" generalizations. For a generalization is law-like, we might say, just in case it is capable of being confirmed by its positive instances.

But even this more limited task quickly runs up against a problem, the "grue paradox" formulated by Goodman. This paradox may be explained as follows. We are asked to suppose that all emeralds examined before a certain time *t* are green. Given that generalizations are supported or confirmed by their positive instances, at time *t* our observations support or confirm the hypothesis that all emeralds are green. "Our evidence statements assert that emerald *a* is green, that emerald *b* is green, and so on; and each confirms the general hypothesis that all emeralds are green. So far, so good."[7]

At this point, Goodman introduces a new predicate, "grue," which "applies to all things examined before *t* just in

[6] In the pair of articles already referred to in the Preface, Wolfgang Stegmüller tries to reconstruct Kant's "solution" along these general lines, by picturing Kant as a "rationalist precursor of the theory of eliminative and enumerative indication." On Stegmüller's reconstruction, the Categories are to be taken as principles of elimination: theories that do not conform to them are ruled out *a priori* as possibly descriptive of our experience, and hence are not confirmable. The fundamental difficulty with Stegmüller's view, apart from the claim that Kant is trying to formulate certain inductive rules, is that on it Kant is trying to prove the validity of Newtonian physics.

[7] *Fact, Fiction, and Forecast* (Indianapolis: The Bobbs-Merrill Company, Inc., 1965), p. 74.

case they are green but to other things just in case they are blue."[8] Now consider the two hypotheses:

H.1: All emeralds are green.
H.2: All emeralds are grue.

It should be clear from the way in which "grue" was introduced that at time *t* all of the evidence for H.1 is also evidence for H.2, and vice-versa. They are equally well confirmed. For at time *t*, the two hypotheses have the same positive instances. However, this is paradoxical. In the first place, although we have been forced to say that they are equally well confirmed, they imply incompatible predictions about emeralds subsequently examined. The fact that all emeralds examined so far have been green—and hence also grue—seems not in the least to support the prediction that the next emerald examined (after *t*) will be grue, although it does seem to support the prediction that it will be green. In the second place, "grue" is an arbitrary predicate. We have no more reason for thinking that emeralds examined after time *t* will be blue than for thinking that they will be red. Hence we have no more reason for thinking that all emeralds are grue than we have for thinking that all are gred. We can cook up any number of "grue"-type predicates. All will be true of emeralds to the same extent that "green" is, for the generalizations in which they figure are supported by precisely the same evidence. But then how are we to distinguish between law-like and accidental generalizations in virtue of the fact that only the former are confirmed by their positive instances? If any series of observations confirms just about any arbitrary generalization, there is no way to distinguish between law-like and accidental generalizations in this respect. Intuitively, H.1 and H.2 are not both law-like; at the very least, H.2 is not. But they are equally well confirmed by their positive instances.

It should be clear how what Goodman calls the "new riddle of induction" is related to Hume's original problem. Hume

[8] *Ibid.*

wanted to know what are the grounds upon which we are entitled to predict the occurrences of future or otherwise unsampled events. Or, to put it in a slightly different way, how are we to distinguish between valid and invalid predictions, that is, between the occasions on which we are entitled to go beyond the present evidence and those on which we are not? Let us say that those predictions are valid which are made in conformity with laws. This is just another way of saying that the validity of particular causal claims depends on the existence of appropriate inductive generalizations. But if this is the case, we must be able to distinguish between laws and non-laws. If we say that laws are true law-like statements, and that law-like statements are statements confirmed by their positive instances—that is, induction "works" or is reasonable with respect to law-like statements—then we run up against the problem posed by the "grue paradox." We could put the same problem in a slightly different way. It appears as though there are too many regularities in our experience, in the sense that to "project" them all to future cases would lead to incompatible results. We need a way of distinguishing between them so as to rule out the non-project*ible*.

Kant's "solution"

Enter Kant![9] We have already mentioned that for Hume the problem of induction is intimately connected with his analysis of causality. For Kant, causality is intimately connected with *his* analysis of the concept of a physical object. Which is to say that Kant's "solution" to the problem of induction draws as much on the first as it does on the second Analogy.

Recall that, given the argument of the first Analogy, events are to be construed as changes in objects, and that on the second Analogy these changes in objects are in every event caused. But to say that these changes are caused is, for Kant, to say that they take place in accordance with some rule or law. The difficulty with "grue"-type predicates is that they are not very well geared to the concept of rule-governed

[9] By way of Davidson. See his note, "Emeroses by Other Names," *Journal of Philosophy*, 63 (1966), pp. 778-780.

change, and as a result not very well geared to the concept of causality either. For something that stays grue changes. Or, to put it the other way around, the change that an emerald undergoes before and after *t* on the "grue" hypothesis is uncaused.

The point needs elaboration. Our problem is to distinguish between

H.1: All emeralds are green,
H.2: All emeralds are grue,

via the notion of law-likeness. But if we construe law-likeness in terms of confirmability via positive instances, there seems to be no way to distinguish between H.1 and H.2; they are equally law-like. And this conflicts with our intuition that H.1, but not H.2, is law-like, that "green" but not "grue" is inductive with respect to emeralds. The source of our intuition seems to be the thought that a change from green to blue as a function of the time at which they are examined is not the type of rule-governed change that emeralds could undergo, perhaps in part because we do not think that time itself is causally efficacious. Hence "grue"-type predicates are not inductive with respect to emeralds. It is this same intuition, generalized, that Kant exploits in the first and second Analogies. The concept of a cause is intimately linked to the concept of an object, by way of an analysis of the concept of change.

This is not to say that there are not possible objects for which "grue"-type predicates are appropriate. "Grue" is, for example, inductive with respect to *emerires*. Something is an emerire if it is examined before *t* and is an emerald, and otherwise is a sapphire. Moreover, the hypothesis "All emerires are grue" is entailed by the conjunction of law-like generalizations "All emeralds are green" and "All sapphires are blue." But of course to say this is just to indicate the difference between emeralds and emerires. To use a rather well-entrenched expression, emeralds but not emerires are *natural kinds*.

There are, in fact, two different points here. They must be

distinguished, although I also think that there is a close connection between them. One is that which generalizations are law-like cannot be determined simply by looking at the predicates. Predicates are not inductive (or "projectible") or non-inductive[10] *tout court*, but rather with respect to *types of objects*. If you know anything at all about emeralds, then you know very generally what sorts of changes they can and cannot undergo. Clearly an emerald cannot undergo a "grue"-type change, for once again such a change would not be rule-governed; that is, it would be tantamount to being uncaused. Just as clearly, emerires can undergo a "grue"-type change, for the color change is associated in a rule-governed way with other changes in the object, i.e., at time *t* emerires undergo a "substantial" change. Of course, it is not part of knowing what emeralds are to know that they are green. But it is part of knowing what they are to know that they have certain

[10] Given a particular analysis of the concept of matter, however, we could begin to classify predicates in just such a way. In fact, two systems of classification suggest themselves. One follows our earlier distinction between objects (or their corresponding concepts) according to their "possibility." Thus we would have "impossible," "merely possible," and "really possible" predicates. "Impossible" predicates would be those either self-contradictory or incapable of being given in experience; "merely possible" predicates those which could be given in experience but which were incapable of being "constructed." On this classification, projectability would be in large part a function of mathematizability and "real changes" in an object would be tied to the conditions of the possibility of experience (recall Kant's remark in the *MFNS* that uniform, rectilinear motion is a "merely possible predicate" of objects). The second system of classification has more narrowly to do with a "subjective"/"objective" contrast. For example, in the course of characterizing his distinction between judgments of perception and judgments of experience in the *Prolegomena*, Kant comments on the projectability of so-called secondary quality predicates. After saying that "That the room is warm, sugar is sweet, wormwood is nasty, are merely subjective judgments," he adds in a footnote: "I readily admit that these examples do not represent such judgments of perception as ever could become judgments of experience, even if a concept of the understanding were added, because they refer merely to feeling, which everyone recognizes as subjective and which one can never attribute to the object, and thus they can never become objective" (#19). Again on this system of classification, the projectability of predicates would seem to depend on how "objective experience" was to be characterized.

causal properties (e.g., to reflect the appropriate light waves), which causal properties in turn are connected with the sorts of changes emeralds can undergo.

The second point is that emeralds but not emerires are a *natural kind.* The idea of a natural kind is so basic, the concept of similarity with which it is linked is so intuitive, that it resists analysis. But emeralds are clearly a kind of thing, as are men and ravens. By the same token, emerires are not a kind, for an emerire examined before *t* and an emerire examined after *t* have nothing (in particular) in common, there is no (relevant) respect in which they are similar. For the same reason, we could say that "green" is a *kind-predicate*, while "grue" is not. If emerires *were* a kind (in which case we would have to know more about them than we do at present), then "grue" would be a kind-predicate. But emerires are not a kind. If we then lay it down that projectible predicates must be kind-predicates, then "grue" will not be projectible.[11]

It has been objected,[12] however, that there is at least one way in which this solution to Goodman's riddle begs the question. For "green" can similarly be defined in terms of "grue" and "bleen."[13] But then, for someone to whom "grue" and "bleen" were primitive color predicates, an object that *stayed green* would undergo an uncaused (unaccompanied) change, from grue to bleen! So "green" is no more, and no less, suited to the rule-governed sorts of changes that emeralds undergo than is "grue," and hence no more projectible with respect to them. To meet this objection, we need to make use of an idea that is an important element in Kant's general account of law-likeness, but that I have not emphasized: the idea of the systematic interconnection of empirical laws as a presupposition of scientific activity. In the Intro-

[11] For more on the connection between projectability and kinds, see W. V. Quine, "Natural Kinds," in *Ontological Relativity* (New York: Columbia University Press, 1969).

[12] In conversation, by Bas van Fraassen and, independently, John Moreland.

[13] An object is bleen just in case it is blue up to some specified time *t*, and otherwise green.

duction to the *Critique of Judgment,* Kant states this idea as follows:

"We must . . . think in nature, in respect of its merely empirical laws, a possibility of infinitely various empirical laws which are, as far as our insight goes, contingent (cannot be cognized *a priori*) and in respect of which we judge nature, according to empirical laws and the possibility of the unity of experience (as a system according to empirical laws), to be contingent. But such a unity must necessarily be presupposed and assumed, for otherwise there would be no thoroughgoing connection of empirical cognitions in a whole of experience."[14]

The question then becomes: can generalizations that take "grue" and "bleen" as primitive be as successfully integrated into a *system of laws* as generalizations that take "green" as primitive? And the answer, I think, must be no. The reason has, indirectly, already been indicated. The projection of "grue" would entail systematic changes. To say that an object has a particular color——— is to say that it has certain causal properties in virtue of which, under specifiable conditions and to normal perceivers, it would appear———. Among such properties would be the capacity to reflect light of particular wave lengths, etc., a dispositional property that we assume can, in turn, be given an explanation in terms of other, non-dispositional, properties. The projection of "grue" would necessarily bring about some changes in this account, at the first stage, for example, in our correlation of colors with particular wave lengths, eventually with the way in which we explained how the wave lengths associated with "grue" were themselves related to an object's micro-physical constitution. The point is not that we could not, under any circumstances, project "grue." The point is that such projection would necessitate wide-scale changes, if the ideal of unity were to be served, in the system of laws in which "grue"-generalizations came to be embedded,[15] if not also in our concepts of types of objects.

[14] From the Bernard translation, pp. 19-20.

[15] This is not nearly so clear, I think, in the case of the so-called secondary

One way of getting clearer about the notion of a "type of object" is with reference to theoretical contexts. Typically, objects are characterized by the particular theories in which they figure,[16] and hence the range of predicates inductive with respect to them is determined by the relevant theory. For instance, given a certain formulation of the atomic theory of matter, there are some things we can and cannot find out about either atoms or matter. One cannot even ask what color they are, or how they taste, or smell. On the other hand, one *can* ask questions about their position, velocity, and mass. "Red" is not projectible with respect to atoms; "has a mass of *n* milligrams" is. The point is that we can know in advance, *a priori*, what sorts of predicates are inductive with respect to what sorts of objects.[17] For the theory tells us, before we begin to put the question to nature, what its objects and hence what its range of inductive predicates must be like.

At the same time, to say that the inductiveness of predicates is relative to a particular theoretical context is not to say that those predicates which are inductive are determined solely by the primitive vocabulary of the theory. It is not the case, for example, that motion can be attributed to corpuscles simply because certain predicates are primitive in a given formulation of the corpuscular theory. Similarly, "grue" can be defined in terms of predicates we assume to be in the primitive vocabu-

---

qualities, since it is part of what is usually meant by calling them "secondary" (or "subjective") that they are not very extensively systematically interconnected. An attempt to project the predicate "sive" (an object is sive just in case it is six-pointed before *t* and five-pointed after) and the generalization "All snowflakes are sive" would involve a greater number of causal properties.

[16] The word "theory" does not have to be pressed very far here. There are "theories" embodied in ordinary language about persons, physical objects, numbers, etc. I do not mean to imply, moreover, that different theories always imply a difference in type of object.

[17] Of course, we do not know in advance which of our generalizations are true and which false. The type of object, or theory in terms of which type is specified, determines the range of law-like generalizations, not the range of the true law-like generalizations. To say this is, further, not to commit oneself to the view that what there is reduces to which theory about what there is is correct.

lary of a theory about gems; that is, "is green before *t*," "is blue after *t*," etc. It is other sorts of considerations that rule out "grue"-type predicates.

Suppose we define a new predicate, "Q," in terms of two predicates of Newtonian physics: "Q" is true of some object *a* just in case *a* has a determinate velocity at time *t* or *a* has a determinate position at time *t*, but not both. "Q" can be defined within classical physics—we have just done it. But it is not inductive with respect to the objects whose behavior classical physics describes. Why not? Because, as construed in classical physics it is not possible that the objects studied by that particular theory do not have both a determinate position and velocity simultaneously. Quantum theory, in projecting "Q," at the same time reconstrues the objects that the earlier theory studies.[18]

objects and causes

The intimate connection between the concept of an object and the concept of cause that Kant tries to establish in the first and second Analogies, and that I have been trying to illustrate and apply with respect to Goodman's "new riddle of induction," has a number of aspects. One of these has to do with the permanence and re-identification of the objects of our experience, hence with the possibility of *objective* experience. Strawson, whose reading of the first and second Analogies in many ways agrees with my own, emphasizes this aspect of the argument:

". . . our concepts of objects, and the criteria of reidentification which they embody, must allow for changes in the objective world subject to the limitation that change must be consistent with the possibility of applying those concepts and

[18] Compare Reichenbach's comment on the advent of the theory of relativity, *The Theory of Relativity and A Priori Knowledge*, p. 97: "We notice the change in the concept of object: what was formerly a property of *things* becomes now a property of things and their systems of reference." As I will try to show in the final section of this chapter, we need to be somewhat precise about the respects in which the quantum theory forces this change. Lambert argues in "Logical Truth and Microphysics," for example, that the classical notion of an object need not be given up in the face of certain pressures generated by that theory.

criteria in experience. How is this requirement satisfied? The answer seems to lie in the fact that our concepts of objects are linked with sets of conditional expectations about the kinds of things we perceive as falling under them. For every kind of object, we can draw up lists of ways in which we shall expect it not to change unless . . . , lists of ways in which we shall expect it to change unless . . . , and lists of ways in which we shall expect it to change if . . . , where, with respect to every type of change or non-change listed, the subordinate clauses introduce further and indefinite lists of clauses each of which would constitute an explanatory condition of the change or absence of change in question."[19]

The re-identification of objects that must be possible if we are to be able to reconstrue them as existing and persisting independently of our perception of them—hence if we are to construe them in Kant's sense as *objects* of experience—depends on the changes that objects undergo being law-like. But it is not simply that objects are, of necessity, entities that behave in a law-like way, entities with respect to which, therefore, induction "works." It is also the case that the identification of a type or kind of object involves the fact that objects of the type or kind behave in particular ways. Strawson puts the point by saying that our concepts of objects are linked with sets of conditional expectations about the things we perceive as falling under them. What he means is that if we know what the objects are like we know not only that laws apply to their behavior, but more particularly whether given types of predicates apply to them. We cannot know in advance every law that holds of every type of object. Appeal must be made to experience. But, on the other hand, the question, what sorts of predicates go together with what sorts of objects—that is, what generalizations are law-like, if not also laws—must be an *a priori* matter, for otherwise identification and re-identification of these objects is out of the question.[20] This is, among other things, what serves to distin-

[19] *The Bounds of Sense*, p. 145.

[20] This seems to be one of the points T. S. Kuhn wants to make in *The Structure of Scientific Revolutions* (Chicago: University of Chicago Press, 1962). In his terms, the very conduct of "normal science" depends on certain concepts and methods having already been established as paradigmatic.

guish physical objects and sense-data, that identification of the former but not of the latter, depends on the fact that particular causal regularities obtain in their behavior. Or, to put it in another, more Lockean, way, the concept of an object is, importantly, the concept of that which is disposed to act in certain ways.[21] And to ascribe a disposition to an object is, once again, to ascribe conditional expectations or law-like regularities in its behavior to it. It is primarily this point that Kant wants to make at A207/B252 in commenting on the proof of the second Analogy:

"How anything can be altered, and how it should be possible that upon one state in a given moment an opposite state may follow in the next moment—of this we have not, *a priori*, the least conception. For that we require knowledge of actual forces, which can only be given empirically as, for instance, of the moving forces, or what amounts to the same thing, of certain successive appearances, as motions, which indicate (the presence of) such forces. But apart from all question of what the content of the alteration, that is, what the state which is altered, may be, the form of every alteration, the condition under which, as a coming to be of another state, it can alone take place, and so the succession of states themselves (the happening), can still be considered *a priori* according to the law of causality and the conditions of time."

Hume revisited

These same points can be made in connection with a closer examination of Hume's position. A closer examination of that position should also help us to appreciate more fully the character of Kant's "solution" to the problem of induction.

There is a great deal of slippage in Hume's discussion of causality in the *Treatise* between taking changes as changes in perceptions and as changes in objects, between taking causes and effects as perceptions and as events. But, one might argue, Hume reaches the conclusion that he does only by way of such slippage. For if, following Kant, we construe changes

[21] By way of contrast, sense-data do not have dispositions.

as changes in or to objects,[22] it follows that such changes are rule-governed or caused. For it is part of our concept of physical objects generally that they do not change without cause and part of our concepts of particular objects that they change only in certain ways and with respect to particular types of causes. Kant tries to establish this conceptual connection in the first and second Analogies, by considering the possibility of objective time-determination. I have tried to support it here by considering the possibility of framing law-like generalizations about objects, and the criteria of identification for them. The point about identification might also be put by saying that the objectivity of physical objects entails that the changes they undergo are rule-governed.

Two passages in the *Treatise* are of particular importance for my purposes. The first is in section VI ("Of modes and substances") of Part I of Book I (page 16 of the Selby-Bigge edition).

"The idea of a substance as well as that of a mode, is nothing but a collection of simple ideas, that are united by the imagination, and have a particular name assigned them, by which we are able to recall, either to ourselves or others, that collection. But the difference betwixt these ideas consists in this, that the particular qualities, which form a substance, are commonly refer'd to an unknown *something*, in which they are supposed to inhere; or granting this fiction should not take place, are at least supposed to be closely and inseparably connected by the relations of contiguity and causality. The effect of this is, that whatever new simple quality we discover to have the same connection with the rest, we immediately comprehend it among them, even 'tho it did not enter into the first conception of the substance. Thus our idea of gold may at first be a yellow colour, malleableness, fusibility; but upon the discovery of its dissolubility in *aqua regia*, we join that to the other qualities, and suppose it to belong to the substance as much as if its idea had from the beginning made part of the compound one. The principle of union being regarded as the

[22] Note that most of Hume's discussion of causes and changes is with respect to objects, billiard balls for example, and not perceptions.

chief part of the complex idea, gives entrance to whatever quality afterward occurs, and is equally comprehended by it, as are the others, which first presented themselves."

I want to draw particular attention to three points in this passage. The first is that, according to Hume, dispositional properties, for example, dissolubility in *aqua regia*, form part of our concept of physical objects, for example, gold. In this he seems to follow Locke, who remarks[23] that "powers"—by which he, Locke, means dispositional properties generally —make up a great part of our idea of substances.

But, and this is the second point, dispositional properties cannot be ascribed to perceptions, only to objects. This is, of course, a claim Hume does not want to make. But even his way of setting out the example betrays him. "Thus our idea of gold may at first be a yellow colour, weight, malleability, fusibility; but upon the discovery of its dissolubility in *aqua regia*, we join that to the other qualities, and suppose it to belong to the substance as much as if its idea had from the very beginning made part of the compound one." The crucial word here is "its." The antecedent of the pronoun can only be the gold. We do not discover the dissolubility of a perception or of a bundle of perceptions in *aqua regia*. Perceptions and bundles of perceptions do not have dispositions, physical objects do. But to say that physical objects have dispositions is just to say that their behavior is law-like or rule-governed.[24]

The third point is that, according to Hume, we discover from experience that gold is dissoluble in *aqua regia*. For Hume, presumably this is true of all the properties we predicate of gold. We learn from experience that gold is yellow, malleable, fusible.[25] But how can we learn all this from expe-

[23] *Essay Concerning Human Understanding*, Book II, chapter 23.

[24] I realize that this is to take a very short way with Hume; the argument needs elaboration. Much has been made of precisely the same point in recent literature in connection with the difficulties phenomenalism faces in analysing subjunctive conditionals. See, for example, Wilfrid Sellars' paper, "Phenomenalism" in *Science, Perception, and Reality*.

[25] Notice, by the way, that these are all (controversially, perhaps, in the case of yellow) dispositional properties.

rience if we do not know what gold is to begin with? Unless we have first identified a substance as *gold*, it makes little sense to predicate anything at all of it. On the other hand, to identify (and to be able to re-identify) some substance as gold is, to use Strawson's expression, to form certain "conditional expectations" about it. But this is just to ascribe certain dispositional properties to it from the outset.

The second passage in the *Treatise* of particular importance for my purposes is in Book I, Part III, section XIV ("Of the idea of necessary connexion").[26] Again, this time without quoting the whole passage, I want to draw attention to three points in it.

First, Hume equates the concept of a *power* with the concept of a *necessary connexion*: "I begin with observing that the terms of *efficacy, agency, power, force, energy, necessity, connexion*, and *productive quality*, are all nearly synonymous; and therefore 'tis an absurdity to employ any of them in defining the rest." I do not think it is a distortion of Hume's position to say that, roughly, this comes to saying that when we ascribe a "power," or dispositional property, to an object, we claim that in a certain respect the object behaves in a rule-governed way. But to say that the object behaves in a rule-governed way is to say that if something happens to the object, for instance, if it is dipped into *aqua regia*, then something else will happen, for instance, the object will dissolve. And this in turn, apparently, is to say that the respective changes in the states of the object are necessarily connected.

The second point is one made in the previously quoted passage, that our ascription of particular powers to objects is on the basis of sense experience alone: ". . . since reason can never give rise to the idea of efficacy, that idea must be deriv'd from experience, and from some particular instances of this efficacy, which make their passage into the mind by the common channels of sensation or reflection. Ideas always represent their objects or impressions; and *vice-versa* there are some objects necessary to give rise to every idea. If we pre-

[26] Pp. 155ff., of the Selby-Bigge edition.

tend, therefore, to have any just idea of this efficacy, we must produce some instance, wherein the efficacy is plainly discoverable to the mind, and its operations obvious to our consciousness or sensation" (Selby-Bigge, pp. 157-158). I will come back to this point in a moment.

The third point takes the second as its premise. It is that power and necessity are qualities of perceptions, not of objects: "The idea of necessity arises from some impression. There is no impression convey'd by our sense, which can give rise to that idea. It must, therefore, be deriv'd from some internal impression, or impression of reflection. There is no internal impression, which has any relation to the present business, but that propensity, which custom produces, to pass from an object to its usual attendent. This therefore is the essence of necessity. Upon the whole, necessity is something that exists in the mind, not in objects; nor is it possible for us ever to form the most distant idea of it, consider'd as a quality in bodies. . . . Thus as the necessity, which makes two times two equal to four, or three angles of a triangle equal to two right ones, lies only in an act of the understanding, by which we consider and compare these ideas; in like manner the necessity or power, which unites causes and effects, lies in the determination of the mind to pass from one to the other. The efficacy or energy of causes is neither plac'd in the causes themselves, nor in the deity, nor in the concurrence of these two principles; but belongs entirely to the soul, which considered the union of two or more objects in all past instances. 'Tis here that the real power of causes is placed, along with their connexion and necessity" (Selby-Bigge, pp. 165-166).

I suggest that Kant's reply to Hume, and ultimately his "solution" to the problem of induction, can be put in the form of a *reductio* of Hume's argument. Objects cannot be "reduced" to bundles of perceptions, as the Transcendental Deduction shows. But the concept of an object as thus irreducible, the concept of an independently existing and persisting object, is in part the concept of that which has dispositional properties, or "powers," as the first and second Analogies show. "Powers" are qualities of objects, not of perceptions. But to say that an object has a power is to say that various of its states are necessarily connected, as Hume himself admits. In brief, Kant's "solution" to the general problem of induction is to show that it is not a problem.[27] Or, rather, that the problem of induction arises only when we construe objects as bundles of perceptions and events and causes as sequences of perceptions. On the contrary, if the concepts of object and cause are "re-instated,"[28] then we know in advance that changes will be law-governed, that induction will "work." If induction does not "work," there will be no *objective* world, hence no unity of consciousness.

Kant and quantum mechanics

It is often said that developments in contemporary physics, particularly quantum mechanics, upset Kant's position in the second Analogy. I will conclude by commenting very briefly on this claim.

There are at least three ways in which quantum mechanics might be thought to pose problems for Kant.[29] All have to do with the alleged "indeterminacy" of quantum mechanics.[30] Indicative of the elusiveness of the issues involved, each seems to turn on a different sense of "indeterminacy."

In the first sense, a theory is said to be "indeterministic" if, given a state-description of some system of objects at an initial time t, the theory does not logically imply a unique state-description of the same system for any other time. It is in just this sense that my reconstruction of Kant's argument in the

[27] At least not for philosophers. Presumably he would admit that there are interesting and important problems for the mathematician working in statistics and probability theory.

[28] Given their intimate connection, they must be "re-instated" together.

[29] A fourth way is suggested by Lambert in "Logical Truth and Microphysics." His solution to the problem posed, which consists in distinguishing between excluded middle and bivalence, is (given our reconstruction of his argument) available to Kant.

[30] In the following discussion, I have drawn on Ernest Nagel's account of causality and indeterminism in chapter 10 of *The Structure of Science*.

second Analogy required that physical laws be "deterministic."[31] If physical laws are not deterministic in this sense, then, I argued on Kant's behalf, we cannot use them to construct a determinate temporal order. It is "determinism" in this same sense that Kant identifies with causal order. But, in this sense, quantum mechanics is not "indeterministic." For given the values of the Psi-function that describes the state of a quantum-mechanical system at some time *t*, the values of the function (for each point of the region over which the function is defined) for all other times are uniquely determined.

In the second sense, a theory is said to be "indeterministic" if the state-descriptions of the system of objects it describes are irreducibly statistical. More accurately, not the Psi-function itself but the square of its amplitude is interpreted as a probability distribution function.[32] In this respect, the sub-atomic processes that quantum mechanical laws explain are taken to be statistical aggregates. As a result, statements attributing properties to *individual* sub-atomic particles are more or less likely, relative to statements concerning the relative frequency with which such properties occur in a class of particles. Kant unquestionably thought of state-descriptions on the model of classical particle mechanics, in terms of the positions and momenta of point-masses. These state-descriptions are non-statistical in character. But I do not think that anything in Kant's argument, as I have reconstructed it, commits him to the claims either that state-descriptions must inevitably be mechanical, i.e., in terms of position and momenta, or non-statistical. Kant did not contemplate either possibility. But neither is it a consequence of his argument, so far as I can see, that a world the basic objects of which were statistically described is not "really possible." It would be a world in which our insight into the behavior of individual objects is limited. And for this reason we might prefer a classical physical world. But it would not seem, in this respect, to be a world of which our experience is not "objective."

In the third sense, "indeterminacy" has to do more directly

[31] See chapter 7.

[32] The probability that basic objects of the system are at various points.

with the Heisenberg uncertainty relations. It is a consequence of these relations that a simultaneous and precise determination of a particle's position and momentum is impossible. It is sometimes suggested, often in connection with the misleading claim that the uncertainty relations derive from the fact that in measuring the movements of sub-atomic particles we inevitably introduce perturbations in these movements,[33] that we cannot *determine* position and momentum precisely, leaving open the possibility that the particles in fact have simultaneously precise positions and momenta. But a more plausible interpretation of the uncertainty relations, perhaps, is that within the context of quantum mechanics statements attributing a precise and simultaneous position and momentum are meaningless.

In this sense of the word, the "indeterminacy" of quantum mechanics *does* appear to undermine Kant's position. For it is part of his insistence on the determinateness of experience that the objects of that experience be determinate with respect to their positions and momenta. It is, I have argued, just the task of the understanding or, from the conceptual point of view, the Categories, to guarantee that the objects of experience be determinate in this respect. A world not determinate in this respect would not be "really possible." The *Metaphysical Foundations of Natural Science* is intended to nail down this part of Kant's argument. But I am not sure that this is the best way to indicate the fatal implications of "indeterminacy" for Kant's position. When Kant talks about "position" and "momentum," he has in mind their characterization in classical mechanics. When "position" and "momentum" are used within the context of quantum theory, it would seem to be the case that they no longer retain their classical meaning.[34] After all, it simply makes no sense in classical mechanics to deny that particles have a precise and simultaneous momentum and position. Perhaps the same thing should be said of "particle" and "wave." Although there are certain analogies between the particles and waves of classical mechanics and of

[33] As if we somehow could not help bumping into the cyclotron.

[34] Again, I follow Nagel, *The Structure of Science*, pp. 297ff.

quantum mechanics,[35] the very fact that the "particles" of quantum mechanics can sometimes and for certain purposes be described as "waves" indicates that we are no longer operating with the classical concept. The conclusion appears inescapable that "momentum," "position," "particle," "wave," have all, in the course of the scientific revolution involved, changed their meanings.[36] But this is just what Kant's position, particularly as exemplified in the *Metaphysical Foundations of Natural Science*, precludes. His position rests in part on the implicit claim that basic concepts are not subject to change, that we can fix permanently the boundaries of the "really possible." But concepts do seem to shift their meanings over time; even in well-defined contexts, truth values are redistributed. To return at last to the theme of this chapter: to characterize inductiveness, or law-likeness, definitively is to assume that we can characterize our concepts of objects and their basic properties definitively. But this is what we cannot do. If there is a general philosophical problem of induction, it is to say how and why and in what respects conceptual shifts take place. It is to be able to say how the kind of scientific progress Kant attempted to rule out on *a priori* grounds is possible.

[35] Formal analogies in particular.

[36] A fact often signaled by putting these words in (scare) quotation-marks.

Selected Bibliography

I. Books

Bird, Graham. *Kant's Theory of Knowledge* (London: Routledge & Kegan Paul, 1962).

Buchdahl, Gerd. *Metaphysics and the Philosophy of Science* (Cambridge: The MIT Press, 1969).

van Fraassen, Bas. *An Introduction to the Philosophy of Space and Time* (New York: Random House, 1970).

Frege, Gottlob. *The Foundations of Arithmetic*, translated by J. L. Austin (New York: Harper & Row, 1953).

Grünbaum, Adolf. *Philosophical Problems of Space and Time*, second edition (Dordrecht: D. Reidel Publishing Company, 1973).

Hintikka, Jaakko. *Logic, Language-Games and Information* (Oxford: Clarendon Press, 1973).

Hoppe, Hansgeorg. *Kants Theorie der Physik* (Frankfurt am Main: Vittorio Klosterman, 1969).

Kant, I. *Critique of Judgment*, translated by J. H. Bernard (New York: Hafner Publishing Co., 1968).

———. *Critique of Pure Reason*, translated by Norman Kemp Smith, second impression with corrections (London: Macmillan and Co., Ltd., 1933).

———. *Logic*, translated by R. S. Hartman and W. Schwarz (New York: The Bobbs-Merrill Company, Inc., 1974).

———. *Metaphysical Foundations of Natural Science*, translated by James Ellington (Indianapolis: The Bobbs-Merrill Company, Inc., 1970).

———. *Prolegomena to Any Future Metaphysics*, translated by Peter Lucas (Manchester: Manchester University Press, 1953).

———. *Selected Pre-Critical Writings*, translated by G. B. Kerferd and D. E. Walford (Manchester: Manchester University Press, 1968).

Martin, Gottfried. *Kant's Metaphysics and Theory of Science* (Manchester: Manchester University Press, 1955).

Mittelstaedt, Peter. *Philosophical Problems of Modern Physics* (Dordrecht: D. Reidel Publishing Company, 1976).

Pap, Arthur. *The A Priori in Physical Theory* (New York: Columbia University Press, 1946).

Plaass, Peter. *Kants Theorie der Naturwissenschaft* (Gottingen: Vandenhoeck & Rupprecht, 1965).

Polonoff, Irving. *Force, Cosmos, Monads and Other Themes of Kant's Early Thought* (Bonn: Bouvier Verlag, 1973).

Reichenbach, Hans. *The Philosophy of Space and Time*, translated by Maria Reichenbach and John Freund (New York: Dover Publications, Inc., 1958).

———. *The Theory of Relativity and A Priori Knowledge* translated by Maria Reichenbach (Berkeley: University of California Press, 1965).

Strawson, P. F. *The Bounds of Sense* (London: Methuen & Co., Ltd., 1966).

Vuillemin, Jules. *Physique et métaphysique kantiennes* (Paris: Presses Universitaires, 1955).

II. Articles

Brittan, Gordon. "Non entis nulla sunt attributa," *Proceedings* of the fourth International Kant Congress (Berlin: Walter de Gruyter, 1974).

Buchdahl, Gerd. "Gravity and Intelligibility: Newton to Kant," in R. E. Butts and J. W. Davis, eds., *The Methodological Heritage of Newton* (Toronto: University of Toronto Press, 1970).

Parsons, Charles. "Kant's Philosophy of Arithmetic," in S. Morgonbesser, P. Suppes, and M. White, eds., *Philosophy, Science and Method* (New York: St. Martin's Press, 1969).

Scholz, H. "Eine Topologie der Zeit im Kantischen Sinne," *Dialectica*, IX (1959).

Stegmüller, Wolfgang. "Towards a Rational Reconstruction of Kant's Metaphysics of Experience," I, *Ratio*, 1967, pp. 1-32, and II, *Ratio*, 1968, pp. 1-37.

Suchting, W. A. "Kant's Second Analogy of Experience," *Kant-Studien*, 1967.

Index

Library of Congress Cataloging in Publication Data

Brittan, Gordon G.
Kant's theory of science.

Includes index.
1. Kant, Immanuel, 1724–1804. 2. Science—Philosophy.
I. Title.
B2799.S3B74 121 77-85531
ISBN 0-691-07221-3